FERDINAND BRAUN

Attendees at the second banquet of the Institute of Radio Engineers, held on 24 April 1915 at Lüchow's in New York City, included many of the ancestors of radio. Ferdinand Braun is marked by (8). Others include (1) G. W. Hayes, (2) J. A. White, (3) J. V. L. Hogan, (4) G. S. DeSousa, (5) E. J. Simon, (6) D. Sarnoff, (7) G. W. Pierce, (9) J. S.

Stone, (10) J. Zenneck, (11) L. de Forest, (12) N. Tesla, (13) F. Loewenstein, (14) A. N. Goldsmith, (15) E. J. Nally, (16) R. A. Weagant, (17) E. F. W. Alexanderson, (18) G. W. Clark, and (19) L. Jones. Photograph courtesy of the Smithsonian Institution, Photo No. 52.269.

FERDINAND BRAUN

A Life of the Nobel Prizewinner and Inventor
of the Cathode-Ray Oscilloscope

Friedrich Kurylo
and
Charles Susskind

The MIT Press
Cambridge, Massachusetts
London, England

Revised edition of Ferdinand Braun: Leben und Wirken des Erfinders der
Braunschen Röhre, Nobelpreis 1909 by Friedrich Kurylo. Originally Pub-
lished in 1965 by Heinz Moos Verlagsgesellschaft mbH & Co. KG,
Munich

This book was set in Penta/202 Bembo by Grafacon, Incorporated, and
printed and bound by Halliday Litho in the United States of America.

Library of Congress Cataloging in Publication Data

Kurylo, Friedrich.

Ferdinand Braun, a life of the Nobel prizewinner and inventor of the
cathode-ray oscilloscope.

"Revised edition of Ferdinand Braun, Leben und Wirken des Erfinders
der Braunschen Röhre, Nobelpreis[träger] 1909"—Verso t.p. Originally
published: Munich: Moos, 1965.

Bibliography: p.

Includes index.

1. Braun, Ferdinand, 1850–1918. 2. Physicists—Germany—Biography. I. Susskind, Charles. II. Title.

QC16.B68K813   530'.092'4 [B]   81-5063
ISBN 0-262-11077-6   AACR2

CONTENTS

6
CREATIVE YEARS: PROFESSOR IN KARLSRUHE, TÜBINGEN, AND STRASBOURG (1883–1918)   48

7
THE FOUNDING OF FIRMS AND GREAT DISCOVERIES (1897–1908)   94

8
YEARS OF HONORS (1909–1918)   179

FOREWORD
Bern Dibner

Ferdinand Braun was certainly one of the great scientists of his time. He discovered the rectifier effect on which all of modern solid-state electronics is based; he made significant contributions to thermodynamics and to the development of magnetic compounds; he invented the cathode-ray oscilloscope, the precursor of the television picture tube; and he advanced radiotelegraphy sufficiently to warrant the joint award of the 1909 Nobel Prize for physics to himself and to Marconi. He was a modest, private man who died an alien in a hostile country in the midst of a world war and has since been unreasonably forgotten by his peers and by historians alike.

Where did Braun come from? One reads that when he was born there was no gas, railroad, water main, or sewer in Fulda, his birthplace in Germany. There was no electricity, and streets were lit by kerosene, homes by oil or candles. The prime mover for Fulda's main industry was a waterwheel. It was not until Braun's graduation from the local *Gymnasium* in 1868 that the railroad was extended to Fulda. He boarded it to reach Marburg, where he attended the university and matriculated in mathematics. In 1869 he moved to the University of Berlin, then the dominant school of science in Germany. His teachers Quincke and Helmholtz sharpened his sense of the essential and of the interplay between theory and experiment.

Braun made his first important discovery while at Leipzig in 1874. He had devised a method of electrical contact with minerals in order to investigate electrolysis. In the course of this work he found that the junctions of some metals and semiconductors did not conform to Ohm's law but were influenced by the magnitude and direction of the current flow. This observation led eventually to the use of semiconductor "rectifiers" in radio crystals, transistors, and solid-state electronic devices. The four papers Braun published in the *Annalen der Physik* on the rectifier effect established the basis for future development.

It is typical of Braun that he made an early decision not to obtain patents for discoveries that resulted from his experimental work. This was an attitude shared by inventors as diverse as Franklin, Faraday, Joseph Henry, Röntgen, and Galileo Ferraris. Franklin's position was that "as we enjoy great advantage from the invention of others, we should be glad of an opportunity to serve others by any invention of ours, and this we should do freely and generously." In July 1898, however, Braun joined two others in a wireless telegraphy patent application so that they would not be limited in further experimentation. Many patents followed.

Braun's career spanned the period of great expansion in Central Europe when Prussia, under Bismarck, seized the initiative to unite the German states into an empire. He was in Berlin in 1871 when Wilhelm I was crowned Emperor of Germany, and his career included appointments at Leipzig, Marburg, and Berlin before he moved to the politically sensitive university at Strasbourg in 1880. His contemporaries and associates included the creators of modern physics—such giants as William Thomson, John Tyndall, Gustav Wiedemann, Hermann Helmholtz, and Conrad Röntgen.

Braun played an important part in the establishment of applied science as an accepted discipline at the universities. At age 33, he undertook a professorship of electrical engineering at the technical university at Karlsruhe in Baden. Studies in the theory of electricity formed part of his work in physics in those early years; it required a broad exposure in the basics of the new science, a knowledge of engineering applications, an ability to pick up new directions, and the confidence to weave these strands into a meaningful academic discipline. At Karlsruhe Braun's contributions included an electrical pyrometer with a reading galvanometer that could be placed at a distance from the furnace, an early example of electrical telemetry. The Braun electrometer, a greatly improved version of the gold-leaf electroscope calibrated in volts, was described in the *Annalen* in 1887. (Heinrich Hertz, who succeeded Braun in

BERN DIBNER

the physics chair at Karlsruhe, was to perform there one of the most significant experiments in electrical history—the demonstration of the existence of electromagnetic waves.)

Braun was one of the early supporters of high-voltage transmission of alternating current. Strasbourg, where Braun taught, was among the first cities in Europe to adopt alternating current at its power station. The power lines were extended to the physics department of the university, and the controversial current became the focus of much investigating and much lecturing. The 50-cycle frequency of the generating station became the standard for most of the world.

The discovery of x rays by Röntgen in 1895 opened a new era in electrical science, as it did in medicine, metallurgy, and communications. The air-exhausted tube, as an instrument for further study, rose in importance and led to "glow tubes" holding gases under pressure, cathode-ray tubes, and later neon tubes and fluorescent lighting. Braun undertook a series of experiments to influence the cathode-ray beam by placing magnets in strategic configurations, causing a spot of moving light on a fluorescent screen at the tube's flared end. This was the origin of the cathode-ray oscilloscope, the television tube, and the computer terminal. The Braun tube was adapted for use in many fields even in Braun's lifetime, especially after a series of improvements and refinements had produced a sharper image. Further work led to radar, the scanning electron microscope, satellite communication, radio-astronomy, and broadcasts to and from space.

As credit for the invention of the electric light goes to Edison not only by virtue of priority but also because of his perseverance in making the device practical and backing it up with a supply system, so credit for the introduction of wireless telegraphy (radio) is given to Marconi in spite of the work of others with Hertzian waves. Marconi was neither a trained physicist nor an electrical engineer, and he therefore faced unresolvable technical problems early in his designs. Braun faced similar problems but drew on his store of physical analogies to build his transmitter for radio-telegraphy. He produced four improvements in circuitry that were recognized by the committee awarding the Nobel Prize. It was also noted that Braun's replacement for the Marconi spark-gap circuit was not simply an improvement but an important new advance. Whereas Marconi's earliest transmitter emitted damped waves with resulting uncertainty of reception, the Braun apparatus generated less damped and therefore more powerful emissions. Braun introduced a sparkless antenna circuit by inserting a magnetic coupler that provided a transformer effect instead of having the antenna directly in the power circuit. He later experimented with methods of improving directional transmission quality; this work was a boon for military applications.

These improvements assured the success of "wireless" and broke the Marconi radiotelegraphy monopoly. Braun was confident that his device could extend the penetrated distance more than fivefold. He had separated the closed oscillating circuit from the coupled antenna circuit and thereby provided an effective and safe apparatus that could extend the transmission range from 15 to over 100 kilometers. In the many experiments aimed at improving and extending transmission range, he reverted to the use of the semiconductor characteristics of certain metal sulfide crystals, with encouraging results. This led to today's transistor. Similarly, Braun's efforts to increase spark-gap length led to the development of powered iron-core inserts into induction coils, coherer improvements, and to today's ferrite core for computer memories and other applications.

The year 1901 was one of triumph for Marconi, who succeeded in spanning the Atlantic from Cornwall to Newfoundland by signaling the letter S; to do this, though, he had to use transmitter and receiver circuits resembling Braun's design. A coalition of competing German radiotelegraph developers was tried but failed, leaving the Marconi firm dominant. The parallel efforts of Marconi and Braun continued for a dozen years, each claiming priority in one form of record or another, each aware of not only the technical features of his improvement but also the political implication of two states heading toward a catastrophic conflict. Parallel developments by the Marconi group and by Braun and other German researchers led inevitably to extended patent litigation to resolve the numerous claims and counterclaims. Under political pressure, the four German radio pioneers pooled their resources to form Telefunken (much as General Electric and Westinghouse were to form RCA).

Braun's election to the coveted position of *Rektor* at the University of Strasbourg reflected his professional abilities and also the importance of the work being done in the advanced laboratories. In his remarks of acceptance of the honor paid him, he spoke of the new atomic age the students faced; radioactivity had been discovered and the physics of the atom was beginning to be gauged; man's flight by airplane had just been realized, and a new dimension in space had been added for man to conquer.

This thrust by theoretical and experimental scientists into the intricacies of the physical world caught the imagination of the advanced nations. The trend encouraged Alfred Nobel to establish his awards, which further stimulated scientific research in the public interest. The first prize in physics went in 1901 to Röntgen. In 1909 it was awarded to Braun and Marconi. The two laureates had approached the solution of "wireless telegraphy" by different routes; both showed the dedication and ingenuity that solved each succeeding problem as the art of radio communication improved. Now, in the presence of Swedish King Gustaf V, each pioneer cordially acknowledged the other's contributions and talents in his prepared papers. It was an occasion for mutual congratulation. A few uncertain years passed and the scene darkened.

Braun was of the opinion that "modern" war was unthinkable among civilized states. He was wrong. When the war came, he withdrew at first from the uncertain border between France and Germany to the quiet of Tübingen, but then returned to Strasbourg as the active front moved west.

Wireless became an important instrument of war. By triangulation ships were spotted and followed, torpedo boats were alerted, and information was relayed. The range of radiotelegraphy increased rapidly from a few hundred kilometers to over 18,000 kilometers, circling half the globe from Germany to New Zealand.

Braun embarked on a Norwegian steamer bound for New York in December 1914 to defend the legitimacy of his work in a patent suit brought against a German-owned radiotelegraph station on Long Island—allegedly an important intelligence link between Germany and America. Marconi himself was expected to appear in court for the prosecution, and therefore Braun was asked to make the trip to appear for the defense. Marconi's failure to appear avoided a court confrontation, but Braun now found himself with little hope of returning home before war's end. He occupied himself by lecturing before science and engineering groups and preparing outlines for future publications, including a projected *Physics for Women*. He lived in a small house in Brooklyn witnessing the uncertain progress of the war, learning of the death of his wife, old friends, and associates, his health deteriorating to the point of total confinement to bed. When the United States declared war on Germany, his new status as an enemy alien further affected his spirit. In this nadir of his fortunes—ill, aging, suspect, and forgotten—this gentle educator, inventor, and benefactor of mankind died in April 1918.

The present work had its origin in a local dispute in the Catholic town of Fulda. In 1955 a proposal was made to name a new school for Ferdinand Braun, the town's most distinguished, if least-known, son. A description of Braun's achievements aroused local pride, and the decision seemed assured until it was revealed that Braun was not Catholic but Protestant, and that what lay in the town cemetery was not Braun's body but his ashes. The resulting dispute in this Catholic stronghold was eventually resolved in Braun's favor, but not before it had caught the attention of a journalist, Friedrich Kurylo, who was sufficiently interested to start researching Braun's life and who eventually produced a biography that was published in Germany in 1965.

The English edition of Kurylo's biography was produced under the guidance of Mrs. Konrad Ferdinand Braun, daughter-in-law of the scientist. It has benefited greatly from the insight of its translator and adaptor, Charles Susskind, who has drawn upon his unmatched knowledge of the history of the science and technology of electronics to produce a book that will be valuable to historians and that also manages to convey something of the spirit of this warm and humane scientist who deserves not to be forgotten.

BERN DIBNER

IN APPRECIATION

Indebtedness for the English version of Friedrich Kurylo's biography of Ferdinand Braun begins with those scholars who scrutinized Kurylo's manuscript and whose interest extended to its translation: Dr. Walter Brill of Siemens-Germany, Dr. Erich Lölhöffel von Loewensprung of Telefunken, and Herr Carl Morgenstern of Hartmann & Braun. A long search in the United States for a translator finally uncovered Dr. Peter Wolff, who for personal reasons could not continue the work. Another search began.

The choice then fell upon Prof. Charles Susskind, who shortened and edited the text and brought the pertinent history of electronics up to date, with particular reference to British and American sources. In the meantime, help came from both sides of the Atlantic: Dr. Alan E. Erickson, Mrs. Leo Fennelly, Dr. Harold Fisher, Dr. Herbert Goetzeler, Miss Mary Grahn, Mr. and Mrs. Robert Merriam, Dr. Peter Stadler, Dr. Klaus Stadler, Herrn Thomas Voller and Wilfried Braun, Frau Christel Glaser, and Dr. Sigfrid von Weiher were unfailingly helpful. Dr. Gabriel Lengyel gave generously of his time and his knowledge of electronics and its history.

Over all hovered two prodding angels: Prof. Dr. Walther Gerlach in Germany and Dr. Bern Dibner in the United States, without whose devotion and insistence this book might never have appeared.

Ruth Emerson Jackman Braun
June 1980

FERDINAND BRAUN

I

CHILDHOOD YEARS (1850–1868)

Ferdinand Braun was born on 6 June 1850 in Fulda, the ancient bishopric and county seat in the Electorate of Hesse-Kassel. The rolling countryside between the Rhine plateau and the Thuringian forest, a Catholic enclave in a Protestant land, is dominated by a hilltop monastery. The Braun family's house at No. 1 Kanal-strasse is situated not far from Fulda's baroque cathedral (figure 1).

Braun's paternal ancestors came from nearby Bad Hersfeld. They were descendants of one Brun whose name is first recorded in 1423 and whose sons and grandsons founded the family wool-weaving business. With the coming of the Industrial Revolution and the arrival of the first mechanical looms from England, the business gradually declined. Ferdinand's father, Johann Conrad Braun, had sought his fortune in the civil service and in 1822 attained his first post as *Repositor*, or junior court clerk, in Eiterfeld. He married Franziska Göhring, the daughter of his superior, the *Aktuar* (clerk of the court) Johann Wunibald Göhring, and started a family. All went well until a minor incident in 1836 that strongly influenced the young family's future.

The judge of the Eiterfeld court, *Amtsrichter* Simon, had asked Göhring and Braun, as well as the court beadle and the prison warder, to close the court for Ascension Day and to join him for a day's trip to a favorite scenic spot, the nearby hilltop ruin of Stoppelsburg. In vain did Simon try to justify this excursion as an official trip when he was called to account before a commission of inquiry two weeks later. Both Simon and Göhring were suspended from the civil service. The case dragged on for six years.

Figure 1
House where Ferdinand Braun was born (photographed in 1922).

Finally, in 1842, both men were obliged to retire prematurely. Because the commission reasoned that *Repositor* Braun was bound to follow his superiors on the excursion, he was fined only 5 talers for his part in the incident. Furthermore, the commission found that Braun had carried on alone at Eiterfeld "with diligence and aptitude." Therefore, he was promoted to *Aktuar*—his father-in-law's rank—and transferred to Fulda, where he arrived with Franziska and their three oldest boys on 18 May 1843, seven years before Ferdinand's birth.[302],*

Franziska Göhring was the descendant of an old family of sextons from the Catholic town of Nesselried in Baden. Her grandfather, Georg Göhring, had come to Fulda as a lay brother and verger of the cathedral.[119] When Franziska married a Protestant, it was understood that sons would be raised in the father's religion and daughters in the mother's, as was the general custom. They had seven children altogether, five boys and two girls. Ferdinand was their sixth child and youngest son, born in the twelfth year of their marriage, when Franziska was 33 and her husband, 51.

At the time of Ferdinand's birth, Fulda had just expanded beyond its medieval walls. A woolen-goods factory employing 250 workers was the largest source of employment. Its steam engine was the boast of the district. There was no railroad, and travel to Frankfurt, Eisenach, and Würzburg was by horsedrawn coaches. Most streets were unpaved. There were neither water mains nor sewers. Gutters served as drains. There were no gas works and no electricity. The streets were lit by oil lamps, the houses by oil, kerosene, or candles. The prime mover of most of Fulda's industry was the waterwheel.[148]

The year of Ferdinand's birth, only two years after the failure of the 1848 revolution, was not a tranquil one in Hesse-Kassel. The elector, Frederick William I, sought to abrogate the rather liberal constitution that his father had signed in 1831. In the German Confederation, he sided with conservative Austria rather than with his neighbor, Prussia. When he dissolved parliament in 1850, his own civil and military officers openly resisted, and he had to appeal to the confederation to restore order in his own land. Austria

---

*Superscript numbers refer to entries in the bibliography at the back of the book.

and Bavaria sent troops, but Prussia also intervened. The two sides fought a minor skirmish just south of Fulda on 8 November 1850.[154] But it suited neither Austria nor Prussia to go to war just then and break up the German Confederation; peace was restored by a quick treaty.

The Prussians thereupon withdrew from Hesse-Kassel while the Bavarians stayed on. Fulda had to quarter a particularly large contingent of these troops. (It had become usual in the elector's service to transfer anyone suspected of "revolutionary tendencies" to Fulda, which became notorious as the "Hessian Siberia."[148]) At least two or three of the *Strafbayern* (punitive Bavarians) were billeted with each family. Families of major officials were assigned fifty soldiers; officials of Conrad Braun's rank were assigned ten. For Fulda this was a heavy burden. Bavarian soldiers lived in the houses while the rightful inhabitants were forced to move into sheds and outbuildings. Provisions ran short and the poor ate rats.[148]

When Ferdinand was 5, the family moved to a nearby house owned by the widow Therese Dorn at No. 32 Kanalstrasse. The Braun children were growing up. The oldest, Wunibald, had already left home and was learning to be a merchant. The other three brothers—Ludwig, Philipp, and Adolf—and sister Katharina were still in school. A seventh child, Albertine, was born (figure 2).

After Easter of 1856, Ferdinand attended the Lutheran elementary school in Fulda. It was conducted by the strict cantor Heinrich Hill,[146] mainly for the children of Protestant civil servants.[148] Ferdinand was an alert pupil, but no prodigy. He found mental arithmetic particularly difficult—an early indication of a peculiarity which often amused those around him: he was far more at home with the theory of arithmetic than with its practice.[303]

In 1859 Ferdinand followed his four brothers into Fulda's *Gymnasium* (academic high school and junior college). Putting five boys through secondary school was no mean achievement for a court clerk who also had to provide dowries for two daughters. Conrad Braun's income, which depended on court fees, seldom exceeded 100 talers a month.[302] Out of that he was expected to provide for his own clerks and copyists. There were no typewriters: every

4

Figure 2
Family Braun in 1865; standing (left to right) are Ferdinand, Adolf,
Philipp, Ludwig, Wunibald; sitting are Albertine, the parents, and
Katharina.

document had to be made out by hand. How could he manage?
The ministry of justice grew suspicious that he was augmenting
his income. An official inquiry revealed that he was—by serving
as his own secretary and working day and night.[302]

Frau Braun's meticulous housekeeping also helped. At times
there was not enough food to satisfy the growing children. Franziska's oft repeated injunction became a family byword: "Eat this
and then you'll be full!"—an imperative against which there was
little recourse.

Like his brothers, Ferdinand was uncommonly diligent.[158] He
passed the *Gymnasium* entrance examination without difficulty.[154]
The requirements included reading and writing in German and
Latin script, the accurate retelling of a short story, four arithmetic
problems with whole numbers, and knowledge of biblical history.
Ferdinand soon proved to be outstanding in mathematics, natural
sciences, and composition. Yet he was no "grind."[130] He did not
neglect games and sports although his slight stature put him at

a disadvantage. His nickname was *Dee Bopp* (the doll), but his talent elicited wide respect.[130]

The strongest influence on Ferdinand in his school years was Dr. Wilhelm Gies, an excellent teacher of natural sciences who had the reputation of being very demanding but whose ability to detect talent was also well recognized. Among many other scholars, two successive professors of physics at the University of Marburg, Wilhelm Schell and Franz Melde, had been his pupils. The talented Ferdinand was a favorite of Gies. As early as three years before Braun's graduation, Gies advised him to study mathematics in Marburg. Ferdinand, on his part, quickly took to Gies and sought to please him in every way, including adopting his passion for crystallography, which Gies saw as the clearest example of the harmony between the laws of nature and the laws of mathematics.[158] In Ferdinand's essay "Water," written when he was 14, these ideas are clearly reflected in the section on the crystallization of water by freezing.[1] During the following summer (1865), the energetic youth composed an entire textbook on crystallography, complete with some 200 drawings in his own hand.

Gies, to whom the manuscript was submitted, expressed surprise at its quality. He was touched to find in it a detailed plan for teaching crystallography based on one he himself had outlined in a school publication in 1859. The manuscript was shown to the professor of mineralogy in Giessen, Adolf Knop, who also judged it favorably. The book was never printed, though, since no publisher would venture to bring out a book by a 15-year-old high-school student, no matter how precocious or well recommended. Braun's love of crystallography found its fulfillment years later in Leipzig when, experimenting with crystals, he discovered the rectification effect.

In the evenings Gies pursued astronomy, and his young admirer did likewise. All of Ferdinand's older brothers had left home, and the attic room facing the street at Kanalstrasse 32 was vacant. He equipped it as his observatory. A telescope was built, and, with the steeple of a nearby church as a reference, he began making nightly observations with his friends. We have no record of any significant results, but we do know that father Braun, on his way home from his overtime work at the courthouse, would often detect tobacco fumes and hear the sounds of animated discussion, not always scientific, coming from the open window.

Gies's influence on Ferdinand extended to philosophy. Besides undergoing conventional religious instruction—he had been confirmed in the Lutheran faith at 14[158]—the young boy acquired a Kantian conception of the world from his teacher. It could not have been easy to profess that view in a school that served to educate many future priests. The study of science had led Dr. Gies to the sort of deism that sees signs of divine creation in all nature. He would say to a student who tended toward atheism, "Deny the existence of God, would you? Then I suggest you study the human ear!"[146]

Ferdinand also sided with his teacher in a dispute that animated the school system over the role of science instruction in secondary schools. Gies thought its importance was not sufficiently recognized.[146] Ferdinand entered the fray with relish in an escapade that nearly cost him his academic career.

Ferdinand's schoolmate Joseph Stieb had written a botanical essay that was accepted by the *Kurhessische Schulzeitung*, a teachers' journal ordinarily concerned with pedagogical matters and evidently in need of science articles.[158] Thereupon Ferdinand, who was one class above Stieb, felt encouraged to submit his own essay on "Water."[130] It was likewise accepted. Unfortunately, Stieb had ended his article with a "few remarks on science instruction in *Gymnasiums*," which he characterized rather forcefully as lagging miles behind the study of dead languages.[157] (Eight years of Latin and four years of Greek were usual.) Such unheard-of impudence from a high-school student earned both youths reprimands (though Ferdinand, who had written on purely scientific matters was innocent) for "unauthorized publication," since neither piece had been formally sanctioned by the school authorities. Stieb was nearly suspended. "Much harder to take than the reprimand," he wrote years later, "was that *Schulzeitung* never paid us a penny for either article."[130]

In any case, Ferdinand's productivity does not seem to have been affected by this incident. Very soon after, on 18 April 1866, the review *Chemisches Central-Blatt* published his article on "A Somewhat Shortened Method of Producing Ammonium Thiocyanate and Some Other Metallic Thiocyanates." In this paper, the youthful author described an improved and more practical method for the production of certain substances of considerable importance in chemical investigations. It proved to be of sufficient interest to be reprinted in a Russian journal in the same year.[1a] And exactly a year later Ferdinand, who in the meantime had turned 16, had another paper published in *Chemisches Central-Blatt*, on a problem whose practical solution eludes us to this day: "Making Sea Water Potable by Chemical Means." He was sufficiently diffident about his process to characterize it as "possibly not realizable in practice; but," he added, "only an actual experiment can determine that, and anyhow the article would still be of some theoretical interest, since it shows that sea water can be made potable in ways other than ordinary distillation."[1b]

In 1866, Austria and Prussia went to war. Austria was decisively beaten after only seven weeks; although she lost no territory of her own, she had to stand by helplessly while Prussia gobbled up Hesse-Kassel and several other provinces. Ferdinand became a Prussian, his father a district court clerk in the Prussian civil service, and the Electoral Hessian *Gymnasium* a Royal Prussian *Gymnasium*.[14a] No other changes were immediately perceptible. The same principal, Dr. Goebel, the strict inquisitor in the Stieb Affair, remained in office and also taught German in the upper classes.[15a] "I gave Ferdinand Braun his head," he wrote later, "with strict objectivity and full regard for his mental individuality."[130]

Did that mean he was aware of Ferdinand's membership in one of the proscribed secret student societies? Proof of membership meant summary dismissal—and for families of modest means such as the Brauns an end to all hopes of higher education, since they could scarcely support a son studying in another city. The future career of the talented boy would be placed in jeopardy. Did Dr. Goebel know and take that into account?

The prohibition of such fraternities at high schools did make a certain amount of sense. The activities of the various militant *Corps* of university students sometimes took up more of the members' time than did their studies. The authorities feared that such "excesses" might be transmitted to the high schools, which would then become hotbeds of activity and training grounds for the *Corps*. Actually, most student societies at the lower level were quite harmless—a moderate consumption of beer was the main activity—and might well have disappeared if they had not been so vigorously hunted down; prohibition lent them glamor in adolescent eyes.[130] Ferdinand did belong to one. He was charter member and from 1866 to 1868 president of Buchonia, one of three secret societies at the Fulda *Gymnasium*. Evidently he got it out of his system, for he never joined a *Corps* after he went on to the university, even though two of his older brothers were members of the prestigious Teutonia. As he was neither against conviviality nor a teetotaler, he would drink a few beers with them now and then and was ultimately elected an honorary Teuton as a young professor.[172]

In 1868, not quite 18 and the youngest of his class, Braun received his diploma with five As (in mathematics, physics, German, history, and religious studies) and three Bs (in Greek, Latin, and French). This excellent record served to excuse him from the dread oral examination.[303] He was the only one of his class to go on to scientific studies. His brother Philipp, just beginning a probationary year as teacher at the *Gymnasium*, participated in the graduation ceremony. For six months, both brothers were at the same institution, one as teacher, the other as student. Philipp taught Greek and Latin, fortunately in the lower classes. He was thus spared the task of examining his brother in precisely those subjects to which Ferdinand's talents were least suited.[158]

2
UNIVERSITY YEARS IN MARBURG AND BERLIN (1868–1872)

In April 1868, after a short Easter vacation, Ferdinand Braun, not yet 18, said goodbye to his family and left Fulda by railroad for the university town of Marburg. He found rooms with the family of a carpenter in one of the tall half-timber houses then typical of the town.

Marburg was about the same size as Fulda but had quite a different aspect. Fulda is surrounded by wide, flat hills; the roofs of Marburg tower picturesquely along the sides of a river valley. Above Fulda is a monastery; above Marburg, a tall secular fortress, the castle of the landgraves. In Fulda the priestly cassock was a prominent part of the daily scene; the streets of Marburg were dominated by the colorful dress of its students.

In 1868 the University of Marburg was quite small, with only 355 students. It occupied an old monastery in which were many tiny cells but no larger rooms that could be used for lectures. A class of even 20 students had to meet in the living quarters of the professor. Few departments were adequately housed; the Mathematics and Physics Institute was an exception, with its own building near the town's hilly center. The path to this spring of scientific knowledge might be steep and difficult, but Ferdinand came well recommended: he was, like the director of the institute, Franz Melde, a former pupil of Dr. Gies.

Ferdinand enrolled in Melde's course in experimental physics and took, in addition, chemistry and mathematics. With this course of studies he provided himself with essentially all the sciences then taught at universities. Ferdinand's father, who understood the value of a secure position, still hoped that his son would become a high-school teacher, then quite a prestigious post in the civil service. Ferdinand was willing, partly because he had no firm alternative plan. Deep in his heart he might have hoped to teach at a university some day, but no way of reaching that goal was in sight. That also may have been why he took so many chemistry courses, including the laboratory course for students specializing in chemistry. He did not attend any of the lectures on theoretical physics given by *Privatdozent* (unsalaried lecturer) Dr. Wilhelm Feussner, doubtless because his excellent preparation in Fulda had carried him past the point covered by Feussner's lectures.[303]

Electricity played a prominent role in the physical sciences of the late nineteenth century. Progress in electricity had given rise to many practical applications: the generation of electrical energy from batteries and dynamos, the beginnings of industrial electrochemistry, the first cable connection between Europe and America. Accordingly, practical considerations were foremost at many universities. All the students, whether aspiring to careers in industry or to careers in teaching, strove to learn all they could about recent developments and to participate in laboratory research, without whose thorough mastery they would not be able to compete with other talented classmates.

Not so in Melde's institute. Melde's main field of interest was acoustics, a field that some critics characterized as a "played-out mine of physics."[168] He spared no expense in his acoustics experiments. He once stretched steel strings from the tower of the institute to the ground as part of a study of the relation of violin bow to violin sound. He discovered that he could get beyond the upper frequency limit of tuning forks by using thin vibrating metal disks, with the help of which he determined the upper limit of human hearing, between 15,000 and 20,000 vibrations per second. But he made no epochal discoveries.

Eager for knowledge, Ferdinand went as far as he could in this direction. (Part of his doctoral dissertation, a study of how the vibration characteristics of a string are affected by its stiffness and anchoring, had its roots in Meldean acoustics.) But soon after he began his studies at Marburg, he realized that his own scientific path was not Melde's. There were also problems of a personal nature. Melde was very musical and as a high-school student had wanted to become an opera singer. His favorite students were those who could participate in his frequent musical evenings.[168] Ferdinand's natural musical talent was less than average, so that a cordial student-teacher relationship could not be developed in this way.[321] Nevertheless, Melde's influence on Ferdinand's career cannot be denied: not only was he to become a brilliant lecturer on acoustics, it is even possible that his major contribution to radiotelegraphy—the utilization of resonance—had its roots in this early work in acoustics.

From Ludwig Carus, the professor of chemistry, Ferdinand learned more about this related discipline than was usual among physicists. His knowledge of chemistry earned him praise a few years later at the examination for qualification as a high-school teacher,[303] and when he became a university professor, he was able to give outstanding lectures on physical chemistry. This interest continued throughout his life, through scientific work in the border area between physics and chemistry and through his membership in the German Chemical Society.

Mathematics gave the least pleasure to scholar Braun. The lectures of Prof. Friedrich Stegmann, a former physician, emphasized fundamentals to the exclusion of the newer, more difficult, and more interesting parts of mathematics.[166] The lectures of *Privatdozent* Carl Alhard von Drach—who later switched from mathematics to art history—gave rise to the most devastating criticism Ferdinand ever voiced about a teacher. "His lectures are completely indigestible," he wrote his parents. "All he does is to dictate notes that only a stenographer could take down."[103]

Small wonder, then, that after only two semesters he began to search for a more congenial academic home. Not surprisingly, he chose Berlin, the highest peak to which any science student in Germany could then aspire. For that, he wrote home, he would give up the "*Gemütlichkeit* and comforts of Marburg," its easy ways, its inexpensive living, and its beautiful scenery." What is more, he would do so at the beginning of the summer semester, traveling against the usual stream of student traffic from the capital to the provinces. He would leave behind the many friends he had made in Marburg and go to Berlin, where he would be a stranger, one student among many. However, financial problems loomed ahead. As we know from his father's meticulous accounts, his means were barely sufficient for Marburg; in Berlin they would be even more limited.

Berlin, where Ferdinand arrived in April 1869, taught him to think on a larger scale. The city had 750,000 inhabitants, as many as the entire province of Hesse. There were 2,000 students at the university, of whom 263 were studying mathematics and science. Berlin was not yet the metropolis that it was later to become. Other German cities were still seats of government. Business and culture were centered in the west and the south. In the sciences, however, Berlin was the undisputed leader, with the best scholars of the day at its university. That was no accident. Berlin's university had been built up deliberately, as a reaction against Napoleon's edict transferring Halle and its university to the kingdom of Westphalia, presumably with the intent of weakening Prussia intellectually.

Ferdinand quickly found quarters in the university district (in what is now East Berlin).[103] After attending several lectures of various professors to find his bearings, he enrolled on 8 May 1869 as a student in mathematics and the natural sciences. Berlin did prove to be much more expensive than Marburg. He found eating ham or Swiss cheese every night tiresome. "A packet of country sausages would be quite welcome," the young man wrote his mother. "I hate the sausages in Berlin. I haven't touched them all winter."

Johann Conrad Braun's attempts to improve the financial situation of his family by applications for scholarships for his elder sons had all remained fruitless, despite strong recommendations from Fulda's lord mayor.[303] By the time Ferdinand's turn came, his father had given up all hope of help from this quarter and did not make an application. Only gradually did the family's circumstances improve, as the older sons started their professional careers.[145] The much traveled oldest son, Wunibald, used his first savings to start an import business in St. Petersburg that soon became known throughout Russia for its supplies of German steel, machines, and chemical products.[236] Ludwig had become a veterinary surgeon in Bremen; Philipp, a regular teacher at the Fulda *Gymnasium*. Whenever possible they all gathered at home at Christ-

Figure 3
Georg Hermann Quincke (1834–1924), Braun's professor and champion in Berlin and Würzburg.

mas to enjoy family life and give each other gifts. Ferdinand was especially attached to his youngest sister Albertine, and his letters home are full of tender concern for "the little one."[103]

Ferdinand was much impressed by the first lectures he attended in Berlin. His great enthusiasm is reflected in his letters home. Instead of a single professor as at Marburg, in Berlin five professors and a *Privatdozent* lectured just on physics. But even here everything was not to his liking. Ferdinand had enjoyed particularly the lectures of Associate Professor Georg Hermann Quincke (then only 36; see figure 3) and soon proposed to him that the work in acoustics begun at Fulda might serve as a dissertation topic.[103]

Quincke gave him every encouragement but had to decline regretfully when Ferdinand requested support for his experiments. Quincke and his colleague Heinrich Wilhelm Dove had no laboratory facilities at their disposal, and they conducted experiments in their apartments as best they could. To eke out a decent salary Dove, though a full professor, also had to lecture at the King Frederick *Gymnasium*, at the military academy, and at a trade school. In going from one institution to another he carried his instruments through the streets in a shopping basket.

To be sure, there was a physics laboratory in Berlin. Its director was the senior member of the science faculty, Prof. Gustav Magnus, who had built up the laboratory at his own expense over several decades and had then sold it to the state. Even then he ruled over it as though it were his private property.[179] Only a select few were given permission to work in this treasury of costly apparatus. Of the 263 science students at the University of Berlin only 4 were so privileged. Ferdinand Braun was one of the privileged. His very first interview with the almighty Magnus had convinced the professor decisively of Ferdinand's talent. It really was a singular distinction: only the most promising students had been admitted, among them some of the future great names in physics, for example, Helmholtz, Tyndall, Wiedemann, and Wüllner.[179] The best equipment in the land was at Braun's disposal, "in complete privacy and gratis," as he reported, and Magnus was the best and most assiduous of teachers. He helped the students set up their experiments and thought nothing of coming in on a Sunday morning to do so. Magnus's lectures were meticulous; his demonstrations were models of their kind. He was the first in Germany to institute physics colloquia at which professors and students could meet informally to discuss the newest scientific developments. Ferdinand flourished in this atmosphere, soaking up knowledge and methodology that would stand him in good stead in his own career as university lecturer.

One man he could not decide to include among his teachers was the second full professor, Dove, who held that no scholar (unless he was another Newton) could hope to hit on more than one truly important idea in a lifetime—not a welcome opinion to a young enthusiast of 19![178] Always bearing in mind that his father's heart was set on a secondary-school teacher's career for his son, Ferdinand also attended the lectures of philosopher Frederick Harms, who tried to reconcile science with idealistic philosophy. In him, Ferdinand found a teacher whose views came close to his own.[181]

Father Braun's insistence on the security of a teacher's career had led him to agree to Ferdinand's transfer to Berlin only on the condition that after two semesters devoted to slaking his thirst for more specific knowledge, he would return to Marburg and complete his studies with the examination for a teacher's credential.[103] Now Ferdinand found he would have to persuade his father, on whom he depended completely for his support, that he would much rather stay in Berlin.

An opportunity to do so arose in December 1869, when Prof. Quincke offered him the assistantship in a second physics laboratory then being established in Berlin in connection with the *Gewerbeakademie,* a technical school that later became part of the Technical University of Berlin. The offer meant that he could bolster his request to remain in Berlin with the news that now he could partially support himself, since the position paid 25 talers a month. To be sure, he probably could have obtained a similar position with Melde in Marburg, but then he again would be tied to a university where "mathematical physics is neglected and the newer and more difficult mathematics either is not taught or is taught by instructors. Moreover, the doctoral examination is no more difficult than the one in Marburg."[103] What really persuaded his father was that the Berlin appointment started in January, whereas any position in Marburg would not be available until the following autumn and would pay less besides.

Despite the prospect of financial independence, his parents' approval was important to Ferdinand as proof of their support for what he wanted to do in life.

There can be no doubt [he wrote them] that Quincke is both socially and scientifically quite a different person from Melde and that I can only benefit from a position here, since Quincke can teach me much more both in theory and in practice. And it will count for something later to have been an assistant to a man like Quincke. . . . I prefer this position even to being assistant to Magnus, since Quincke is more of a mathematical physicist and is much closer to his students.[103]

Ferdinand remained loyal to Quincke all his life. When Quincke celebrated his seventieth birthday in 1904 Ferdinand, then 54, as Quincke's first personal assistant wrote about him and his work in the *Annalen der Physik*.[80] Ferdinand particularly admired and emulated Quincke's ability to do scientific work with relatively simple means. Moreover, Quincke personified the spirit and method of German academic instruction, which came to serve as the model for science education throughout the world. The two men remained friends until Braun's death.[80]

Ferdinand's hours in Quincke's laboratory were nominally from nine to one and from four to six every day. But on three days of the week, when Quincke was away lecturing at the university, Ferdinand's time was his own.[103] Not only the institute's instruments but also Quincke's private papers were at Ferdinand's disposal. The library of the institute was housed in the same building as the laboratory—a fortunate situation for his purposes. Thus he was able to conclude his investigation of the vibrations of strings and to submit it as his doctoral dissertation in 1872 (figure 4). He had the privilege of choosing his examiner in each subject and he confidently chose Hermann Helmholtz for the major subject, physics.

Helmholtz had come to Berlin from Heidelberg to replace Magnus, who had died in 1870. Helmholtz embodied the epitome of the classical period of physics, when natural scientists still had a command of all fields of science and contributed to all of them. He had advanced considerably the study of acoustics with his major work, *On the Sensations of Tone* (1862). Choosing him as an examiner signified that Ferdinand wanted to have his dissertation in acoustics judged by the leading expert of the time.

The dissertation was a genuinely original contribution. Its most important finding was that in rigid strings under tension, for example metal strings of finite cross section (that is, media in which elasticity cannot be considered negligible), higher tones propagate faster than lower ones; high and low tones launched simultaneously along such a string arrive at a distant point at different times. On Quincke's suggestion, Ferdinand added an "analogy" that sought to relate this phenomenon to corresponding phenomena in optics, namely, to the dispersion of visible light by a prism—a branch

Ueber den Einfluſs

von

Steifigkeit, Befestigung und Amplitude auf die Schwingungen von Saiten.

INAUGURAL-DISSERTATION,

ZUR

ERLANGUNG DER DOCTORWÜRDE

VON DER PHILOSOPHISCHEN FACULTÄT

DER

FRIEDRICH-WILHELMS-UNIVERSITÄT ZU BERLIN

GENEHMIGT

UND

ÖFFENTLICH ZU VERTHEIDIGEN

am 23. März 1872

VON

Ferdinand Braun

aus Fulda.

OPPONENTEN:

E. Bessel Hagen, stud. phil.
F. Poske, stud. phil.
A. Vollen, stud. phil.

BERLIN.

DRUCK VON GUSTAV SCHADE (OTTO FRANCKE).
Marienstr. 10.

Figure 4
Title page of Ferdinand Braun's doctoral dissertation.

of physics in which the type of direct measurement he used in his acoustical studies was not yet feasible.[104] Although means were presently found to make such measurements, so that the suggested analogy was no longer needed, Ferdinand's dissertation was considered a significant contribution to the physics of the day and quickly earned him the beginnings of a reputation.

The dissertation is also interesting because it contains the basic elements of his later scientific method: a clear feeling for the essential problem, experimentation with a definite objective in mind, and ingenious experimental arrangements. His future student and later collaborator Jonathan Zenneck wrote in his memoirs, "Braun carried his experiments only as far as was necessary to clarify the relations. He was never one to make physical measurements as precisely as possible for their own sake."[135] The dissertation contains similar statements: "Although the formulas are merely an approximation, the tables show what I think is quite satisfactory agreement . . ."; and when an auxiliary experiment fails, the candidate, unperturbed, remarks, "My efforts unfortunately were not crowned with success, so that I am unable to provide further data on this question for the present." The approach tallied with Helmholtz's view that "sometimes it takes more ingenuity and thought to make an intractable piece of brass or glass do what it is supposed to do than to devise an entire experimental procedure."[228]

Only in experiments of secondary importance did Ferdinand give up when confronted with "technical difficulties." His basic experiments were extremely well thought out and often exhibited great ingenuity.[123] For instance, to record the vibrations of his strings, Braun attached a tiny barb from a swan's feather to the string with a pinpoint of glue and allowed the barb to trace its own pattern on a rotating cylinder of carbon-covered paper arranged alongside the string. The waveforms differed according to whether the string was struck, bowed, or plucked; whether it had been partially filed, heated to a glow, or twisted; and whether it was damped by blocks of wood or iron. Braun spared no effort in his quest to detect the nodes and antinodes of the vibrating string.

Ferdinand Braun, the first Berlin candidate to be examined by Helmholtz, passed his doctoral examination *cum laude* on the evening of 8 February 1872. "On the whole, it went quite well," he reported to his parents a day later. "The ceremony will take place after the dissertation has been printed."[103]

The ceremony was on the morning of Saturday, March 23. After a public defense of his dissertation against three opponents, as was then customary, the 21-year-old Ferdinand Braun was formally granted the degree of doctor of philosophy, though not without some cost. He had had to buy a new black vest and a new hat for the doctoral examination, tip the university servants, and entertain the three opponents. The greatest shock was the cost of printing the dissertation, even though the printed version was restricted to the minimum twenty-four pages required by the regulations. The dissertation was dedicated to his parents, "To offer you [as he wrote them] at least a modest testimonial of my gratitude and filial reverence. May it give you as much pleasure as it has given me annoyance over its cost." A special inscribed copy went to his old teacher, Dr. Gies. On 2 September 1872 a paper based on the dissertation was published in the *Annalen der Physik und Chemie*. For the first time the name Ferdinand Braun appeared in a research journal.

3
ASSISTANT IN WÜRZBURG (1872–1874)

No sooner had the new doctor been confirmed in his rights and privileges than he was faced with a double move. After nearly three years at No. 48 Augustusstrasse he had found new rooms overlooking the Luisenkanal, one of Berlin's waterways, where he felt "as good as new."[103] But only a few days later he learned he would have to move again. Quincke had accepted a call to the chair of physics at the University of Würzburg. He had insisted on reserving the right to choose his own assistant, and he chose Ferdinand Braun.

Quincke's offer had resulted from one of those games of academic musical chairs that starts whenever a professor's chair falls vacant. This time the immediate cause had been the war. Alsace-Lorraine had become part of Germany after the war of 1870–1871, and the old University of Strasbourg in the Alsatian capital was to be enlarged and generally improved, especially in the sciences. The physics chair was offered to the highly promising August Kundt at Würzburg, one of the great experimentalists of the nineteenth century. He had come to Würzburg when he was only 30, just three years before, with his assistant Wilhelm Conrad Röntgen, from the Technical University in Zurich—but the housing for experimental physics at Würzburg was then among the worst in the land.[80] He was only too happy to leave for Strasbourg, where a new physics institute was to be erected. With him went his assistant; and Quincke and Braun were to replace them in Würzburg.

The decision whether to go to Würzburg with Quincke or stay in Berlin was difficult. "Should I start as soon as possible on a career as a high-school teacher, or should I set out on an academic career with a long-range view toward a technical university?" Now that he had finished his studies and his professional life was about to begin, the question had become pressing.

Long ago Quincke had recognized his student's talent. Now he noticed that Braun was worried that high-school teaching might bury his hopes for the academic profession. As early as the winter of 1871–1872 he proposed that after graduation Braun should take the qualifying examination for high-school teaching, but that he should stay on as assistant: "Perhaps you might meanwhile be able to get a university position, or some other favorable opportunity might come up. I am often asked to recommend an able physicist and it would please me if I could help you to a position." Quincke had already introduced his assistant to professors, businessmen, and government representatives of his large and influential circle of acquaintances in Berlin. "For my part, I am not disinclined to accept this proposal," Braun wrote home in March 1872. "All this gives me the assurance that he will find a suitable place for me, I mean of course once I have completed the requisite training and after I have published a few scientific articles."[103]

His parents were convinced, especially after their oldest son, Wunibald, put in a word from St. Petersburg: "By all means let Ferdinand enter upon an academic career; there really seems to be a shortage of physicists and the salaries are improving."[103] On 15 April 1872, Braun and Quincke moved to Würzburg.

Not yet financially secure, Braun scarcely could make ends meet on the usual university assistant's salary of 25 talers a month—and Würzburg proposed to allocate only 16 talers. Quincke protested that Braun's predecessor, young Röntgen, had received 25 talers; but he was married, countered the university. Very well, said Quincke, he would pay the difference himself. The university thereupon agreed to pay the full amount.

Under these circumstances Braun felt compelled to seek a supplementary income. "It should not be difficult," Braun wrote home, "to make up what I need by literary activity."[103] He would write, and he would qualify for the secondary teacher's credential without delay. The examination was rigorous: physics, mineralogy, chemistry, psychology and logic, mathematics, astronomy, botany, earth science, pedagogy and history of philosophy, modern and classical literature, and paleontology. On 24 January 1873 Braun was certified to teach high-school mathematics and physical sciences, and junior-high-school German.[303]

As to writing, had not his *Gymnasium* teacher, Dr. Goebel, admired young Braun's style?[130] With his gift for sarcasm, he had a natural bent for the satirical weeklies, of which the outstanding representative was the *Fliegende Blätter* in Munich.[303] Aiming as always for the top, Braun did get several contributions into the magazine; but it would take a lot of literary detective work to identify them all. Most contributions were signed by pseudonyms; sometimes several writers shared a single *nom de plume*. His most probable pseudonym, judging from style and subject matter, was *Crassus* ("Fatty"),[301] a name under which he was to write later.[216] We have an example of his style in a poem incontrovertibly his. It was so sharply aimed at the officiousness of the emerging Second Reich that even the usually fearless *Fliegende Blätter* turned it down, but part of the manuscript survives.[303] It is a longish poem called "*Die philosophischen Wächter*" ("The Philosopher Guardians"), which ends with a devastating attack on the advocates of the authoritarian state of Plato's *Republic*:

Man hat, o Plato, deinen Staat gemacht bei uns zur halben That. Wärst, Plato, selber du am Ruder, ich glaub', 's wär alles unterm L	Thy State, O Plato—what a laugh— We now enjoy, at least by half. Wert thou the coxwain of our shell, We'd altogether go to h . . . .

Braun's studies were not interrupted by military service. As a graduate of a *Gymnasium* he was exempt from all but the single year expected of officer candidates—and he never even had to serve that. He was not called up until 1873 and was promptly declared unfit for service on account of his severe myopia.

Between his assistant's duties and freelancing Braun had little time for research. His only published work dating from 1873 was a second paper published in the *Annalen der Physik und Chemie* in 1874, based, as was his first paper, on his dissertation: "On Elastic Vibrations of Noninfinitesimal Amplitudes." With that, Braun bade farewell to the interests that went back to his days with Melde. A new field, derived from his association with Quincke, now engaged his attention.

Quincke's lifelong interest was the behavior of liquids in thin tubes (capillarity). At Würzburg he investigated capillary properties of fused substances—some fifty of them.[80] Braun's interest was particularly drawn to fused salts. As he watched salt crystals melting in a crucible, it occurred to him that the resulting liquids might provide the answer to a question "many of the best experimentalists" had asked in vain: How was electricity conducted "electrolytically"?[3]

Once again, Braun aimed high. Electrical theory contrasted metallic conduction with electrolytic conduction. Metallic conduction obeyed Ohm's law, first formulated in 1827, according to which current is proportional to voltage, with the constant of proportionality depending in a simple way on the properties of the material and on its shape: a short piece is a better conductor than a long one; a thick one, better than a thin one. But, despite many attempts, no one had contrived to extend the law to solutions of several materials, for instance, salts. "Our ideas on electrolytic conduction become hypothetical—not to say nonexistent," Braun wrote.

He thought the problem lay in the nature of the electrolyte, usually a complex solution of several salts. Little was known about the conductivity of either the solvents or the salts and less still about their interactions. What if one were to look at pure salts liquefied by heat, without a solvent? Braun eagerly threw himself into the research. There were no further contributions to *Fliegende Blätter*. An "extensive preliminary investigation" yielded "experimental methods that proved flawless." Next, Braun checked the conductivities of a dozen melts, but the results were unexpectedly disappointing. If he had hoped to formulate a Braun's law for electrolytes to set beside Ohm's law, he did not succeed. "Conductivity of electrolytes cannot be correlated with any of their other physical or chemical properties," he conceded.[3] (Actually, as is now known, electrolytes do follow Ohm's law; the deviations observed experimentally are due to effects that occur on the electrode surfaces.)

Braun based two papers on this work. One paper went to Alphons Oppenheim, an associate professor of chemistry in Berlin whom Braun knew and who read the paper at a meeting of the recently formed German Chemical Society before publishing it in its proceedings.[3] The other came out in the *Annalen der Physik und Chemie*.[4] The reported results may have been negative, but the episode had saved Braun from becoming a poetaster and set him back on his proper path as scientist.

After two years with Quincke it was time to admit that no chance for a university post was in prospect. For one thing, Quincke was unable to keep up all his Berlin associations in provincial Würzburg. Braun had accomplished one of his aims—he had matured scientifically and academically under Quincke's tutelage, but—if only for financial reasons—he could not remain in an assistant's position forever; already he had had to resort to private tutoring and satirical verse.[103] It looked as if Braun's father had been right. A long-advertised vacancy for a mathematics and science teacher at the Thomas School, a *Gymnasium* in Leipzig, suddenly looked very good. Braun sent his application for it to the Royal Saxon authorities and was accepted. At last he would be financially independent.

From a scientific viewpoint, leaving Quincke was no hardship. Despite Quincke's great plans for it, the University of Würzburg's Physics Institute continued to make do with a single floor in a building that also housed the departments of mineralogy and geology and their collections. The place was incredibly crowded. "On some mornings I had to move gingerly among the demonstration experiments," wrote Braun, "for fear that I might knock over a piece of apparatus and upset the arrangements."[80]

Nor was Quincke without his quirks.[185] As far as he was concerned, thermodynamics did not exist, and he had no interest in the formulation of the laws of energy, one of the great problems of the day. As it turned out, he did not stay in Würzburg either, but left the following year to replace Robert Kirchhoff in Heidelberg (when Kirchhoff went to Berlin after Dove died in 1874) and remained there the rest of his life.

As often happens, Quincke's efforts in Würzburg bore fruit after he left. Under his successor, the 35-year-old Friedrich Kohlrausch, who came there from his position as associate professor in Strasbourg, Würzburg did get a new physics institute. It was to become famous after Kohlrausch left and Röntgen returned there in 1894, the year before he discovered x rays.

4
TEACHER AT THE THOMAS *GYMNASIUM* IN LEIPZIG (1874–1877)

In the fall of 1874 Braun started the usual six-month probationary period as assistant master in Leipzig, with eight classes in natural sciences and two in mathematics—twenty hours of classes a week. He found it easier than he had expected to turn to the career of high-school teacher. He was convinced it was still possible to become a university professor, and he intended to pursue that goal as best he could. He had chosen to go to Leipzig for practical reasons. In Würzburg, after being admitted to university teaching, "there would have been little hope of earning money from lectures," as student fees were the only income of the unsalaried lecturer.

He took up his teaching with confidence and immediately involved himself in intense scientific activity. The unsatisfactory results of his Würzburg work on fused salts gave him no peace of mind. He turned to the earlier findings of J. W. Hittorf, who had studied the temperature dependence of copper sulfide, a property first discovered by Michael Faraday. Hittorf showed that this dependence was tied in with the transition from electrolytic to metallic conduction.[237]

Braun studied Hittorf's publications thoroughly. Then he started to look into metal sulfides and other substances that he found "interesting because they conduct in the absence of electrolysis, even though they are binary compounds."[96] The main problem of experimental technique was to find a way to clamp the mineral specimen absolutely securely. Pressing it between two metal plates yielded unreliable contacts; so did making contact through drops of mercury poured into holes that had been bored into the mineral.

Gripping it in vises, as Hittorf himself had done, would be satisfactory, but Braun found the method too complicated.[4]

His solution was to use two lengths of silver wire. One was bent into the shape of a ring; the other was folded several times by crimping, which gave it a spring action. The point of the spring wire, when placed against the mineral specimen, pressed it against the ring-shaped wire—no clamps, screws, vises, or drops of mercury were needed.[4] It was an example of the "elegance" with which Braun solved experimental problems.[131] It also proved to be an indispensable precondition for his first major scientific contribution, the discovery of the rectification effect, that is, the fact that the flow of electricity across certain compounds does not obey Ohm's law, but instead depends on the sign and magnitude of the current. This effect underlies the operation of crystal rectifiers, transistors, and the many other devices of solid-state electronics.

The history of physics does not record when and how Braun made this important discovery—when he first noted that the deflection of his Wiedemann compass galvanometer changed substantially as the polarity of the current source was reversed. One can only imagine his surprise, the careful repetitions of the measurements, the replacement of each successive circuit element—all in vain: the phenomenon was a real one; the outcome, the same again and again. It is summarized in the final sentence of his account of the experiment:

I have found that for a large number of natural and artificial metal sulfides of very different configurations, from the most perfectly formed crystals obtainable to quite rough specimens, the resistance varies according to current direction, intensity, and duration. The variations amounted to as much as 30 percent of the full value.[4]

It is likely that Braun first observed the rectifier effect while he was still at Würzburg. Very little time had elapsed between his arrival in Leipzig at the end of September and the date of his report, 23 November 1874.[4] The final experiments may have been made in Leipzig, perhaps with crystals borrowed from the mineral collection of the Thomas *Gymnasium,* perhaps even in his rooms at No. 89 Weststrasse, a short distance from the Thomas *Gymnasium.* What is certain is that Braun's experimental arrangement—the contact between the point of the silver wire and the semiconducting crystal—was precisely the one in which rectification would be unmistakably observed. Braun was very much aware of this fortuitous circumstance. In a paper published several years later he remarked, "The anomalous phenomena occur most readily if at least one electrode is small."[8]

The first report, "On the Conduction of Electrical Currents through Metal Sulfides," appeared in the *Annalen der Physik und Chemie* at the end of 1874.[4] The results were of considerable scientific interest. So not all metallic substances obeyed Ohm's law after all. Nor did resistance necessarily follow the textbook prescription: direct proportionality to the length of a conductor and inverse proportionality to its cross-sectional area.[226] Here was an entire class of substances that behaved according to quite different, as yet unknown rules. The door to a peculiar, mysterious realm had been thrown open.

Practical applications, notably in electronics, were far in the future; here science was considerably ahead of technology. Not for thirty years, until 1904, would it be realized that junctions of certain metals and crystals were ideal "detectors" of radio signals—the "cat's whisker" crystal rectifiers of early radio. (Braun's crimped-wire electrode was duly reinvented in America in 1910 and patented.[332]) The transistor effect was not discovered until 1948 (figure 5).

In Leipzig, it became known that the new mathematics and science master at the Thomas *Gymnasium* was no ordinary high-school teacher but a young scientist, only 24, who already had three scientific publications to his credit. The Protestant *Gymnasium* in Strasbourg offered him a place, but Braun declined.

At Easter 1875 the school appointed Braun to a permanent position with a monthly salary of 58 talers. Three months later he was promoted by two steps "because he has performed useful services in the instruction of natural science."[189] On 1 December 1875, little over a year after starting in Leipzig, he was promoted to the highest grade possible for a young teacher.

His students were enthusiastic about the liveliness of his lessons. At the time he wrote of his approach, "Keep them at it with pencil and paper, compass and ruler—give them no chance to fall asleep or sit with hands on lap."[6] The young teacher was invited to give

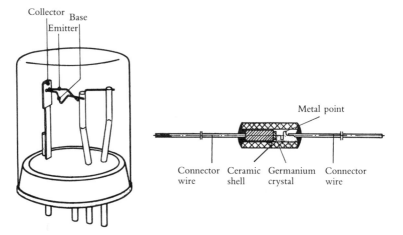

Figure 5
Ferdinand Braun's crimped-wire electrode as a design component of early point-contact semiconductor (left, transistor; right, diode).

lectures to the local science society and also to the general public. He also gave a well-attended adult education course in physics at the Leipzig *Volksbildungsverein*.[103]

The Thomas *Gymnasium*, which traced its existence to a monastery founded in 1213 and which had been made famous through the work of its greatest cantor, Johann Sebastian Bach, was favored by Leipzig's great publisher families.[199] Otto Spamer, a leading publisher of popular science books, soon became aware of the gifted new teacher and invited him to contribute a book on mathematics to his series *New Library for Youth and the Home—Treasures of Knowledge and Information for Young and Old*. The potential author hesitated. Spamer had definite ideas about the sort of book he wanted: it should have a lot of conversation and few equations. Braun doubted that he was the right person to awaken the interest of young readers in mathematics "by means of number games, puzzles, and other drolleries," tricks in which he had little personal interest.[6] In the end, though, the challenge of the project proved irresistible, and he agreed provided its scope was extended to include all of natural science: *Der junge Mathematiker* would become *Der junge Mathematiker und Naturforscher*.

The book appeared in time for the Christmas trade in 1875, selling at 4 marks. It received excellent reviews. The *Frankfurter*

Journal praised the way the author had shown "how numerical rules govern all branches of science."[310] Another Frankfurt paper, the *Intelligenzblatt*, said the book would "surely secure a leading position among the numerous works intended for the edification of the young."[331] The book consisted of 24 "entertainments" prepared by a father for his children, in fortnightly installments spread over the whole year. A little girl reminiscent of Braun's beloved sister Albertine was part of the group. The summer months led naturally to questions about botany and electrical storms; clear autumn nights to discussions about astronomy; and winter days to subjects such as ice and temperature. There were over 1000 exercises, including puzzles, experiments, things to draw, card tricks, games, and reflections on natural philosophy, and some 320 illustrations. "I might be reproached for expecting too much of the reader, a reproach I gladly accept," wrote Braun in his preface. "Perhaps I shall interest only a modest portion of my potential readers, but—and this is my intention—it will be the *better* portion!"—a sentiment characteristic of his own teacher, Wilhelm Gies.

Der junge Mathematiker und Naturforscher was Braun's only book; his scientific work usually took the form of articles or, at most, pamphlets. He was quite pleased with the results, though we have no indication of the commercial success of the book. A somewhat thinner second edition was published in 1881 at a price of $4\frac{1}{2}$ marks[322] (figures 6 and 7).

The lack of a theory,[5] the author's youth, and the absence of practical applications all worked against Braun's paper on conduction in sulfides, so that it failed to make the impression it would have made under different circumstances. He kept on with his experiments, knowing full well how difficult they were, mainly "out of curiosity—a human failing that usually grows worse the longer it remains unsatisfied."[96] The requisite technique was so exacting that few undertook to repeat the observations. The French mineralogist Henri Dufet tried and failed; he declared that Braun was all wrong.[232] Arthur Schuster, then a 23-year-old assistant to James Clerk Maxwell in the Cavendish Laboratory at Cambridge, got results for contacts between clean and tarnished (that is, oxidized) copper wires similar to those Braun had obtained for metals

Figure 6
Title page of *Der junge Mathematiker und Naturforscher*, Braun's 1875 book of "edification and enlightenment for young and old."

1876–1877

Figure 7
Illustration for "an evening's entertainment" from Braun's *Der junge Mathematiker und Naturforscher.*

and sulfides.[333] W. G. Adams and R. E. Day in England[334] and Werner Siemens in Germany[335] worked on the light-sensitive electrical properties of selenium, and likewise found it hard to get unequivocal, reproducible results.

Two years after the first publication Braun read a paper on his subsequent investigations, again before the *Naturforschende Gesellschaft* in Leipzig, under a more restrictive heading: "Experiments on Departures from Ohm's Law in Metallic Conductors." He explained why Dufet's experiments had been unsuccessful: the contacts had been of equal size, whereas one of them should have been a fine point—a detail Braun admittedly had not stressed in his first paper. He also noted that he had obtained some of his results with manganese brown, a mineral outside the sulfide group. But an all-encompassing theory eluded his grasp. He made some pretty shrewd guesses. He had come to the conclusion that the whole phenomenon must take place within a thin surface layer. He had found that the current in the point electrode remained the same whether it was made to flow across the crystal to a single blunt electrode or to a pair of them, whereas it should have dropped to half the value with two electrodes if the entire volume of the crystal had been involved.[8] He sought an analogy with conduction through gases, which was known to depend on the polarity of the current under certain conditions. Even that investigation, reported in a new paper in mid-1875, failed to solve the mystery.[5] The most he could do was to reiterate his previous conviction that the findings were genuine, not artifacts of the experiment or second-order effects such as might have been caused by heating, as Werner Siemens continued to believe.[96] With characteristic ingenuity, Braun proved that particular point by means of a tiny, extremely fast-acting switch. He was able to show that rectification took place even when the current was allowed to flow for only 1/500 second—too short a duration, he thought, to permit thermal effects to set in. Thus, although the theory of semiconductor junctions proved to be beyond him—and small wonder, when we consider that the work was done decades before the discovery of the electron—Braun nevertheless had established the two related characteristics that were to prove of prime technological importance in the distant future: the phenomena took place in a thin surface layer, and they took place rapidly.

There is no evidence that these investigations made a particular impression on Leipzig's two leading physical scientists. One was Gustav Wiedemann, professor of physical chemistry, a former student of Magnus, author of the great *Handbook of Galvanism* and future editor of the *Annalen der Physik und Chemie*. The other was Wilhelm Hankel, professor of physics and acknowledged expert on the properties of crystals, who had studied their electrical characteristics for years without once making a discovery as important as the one the 24-year-old Braun had made on his first try. Both men were present at Braun's lecture before the *Naturforschende Gesellschaft*. Braun had saved a surprise for the end of his lecture: five demonstrations of departures from Ohm's law by means of measurements on manganese brown and galena. The minutes record soberly that all five demonstrations were successful: "The deflection differed according to the direction of the current."[233]

During 1876 Braun also devoted some time, as a follow-up to his doctoral work, to the theory of elasticity. On 13 June 1876, he gave still another lecture to the same audience: "On the Nature of Elastic Recovery."[7] This effect, first discovered by Wilhelm Weber (another physicist better known for his work in electricity) in 1835, occurs in some materials when they are permanently deformed by being pushed beyond their elastic limits. After a time such a material creeps slowly back to a new "set" partway toward its original state. The problem had occupied several physicists; for instance, Braun's contemporary Emil Warburg proposed a solution involving a theory based on the rotation of molecules. Braun's work provided the "experimental foundation" for that theory.[297] Among other things, Braun showed that the final state of a body deformed by two independent forces differed according to whether the forces had been applied simultaneously or consecutively.[7]

The nonlinearity in the stress-strain relation that these effects demonstrate was probably also the cause of the amplitude dependence of the propagation velocities along strings that Braun had described in his dissertation.[2] Was he aware that he was looking at two aspects of the same problem? We cannot answer with absolute certainty, but it seems likely that he was.

5
ASSOCIATE PROFESSOR IN MARBURG AND STRASBOURG (1877–1882)

In the five years following Braun's 1870 departure from Marburg, the state of physics instruction there had gone from bad to worse. Melde had neglected electricity. Even if he had wanted to, he could not have spent the time to teach it well since his was a one-man department. As the number of students who wanted to study electricity increased, the situation became intolerable. In 1875 the academic senate of the university formally requested the Prussian Ministry of Education to establish an associate professorship in physics.

Berlin knew well on whose behalf this request was made. Wilhelm Feussner, a diligent 42-year-old practitioner of routine teaching and a friend of Melde's, well deserved an improvement in his income as well as offical recognition. At least that was what Feussner's friends in Marburg thought. Yet Berlin had different ideas. It was generally known that Feussner's lectures were quite old fashioned, and Berlin did not want students skipping courses on theoretical physics, as Braun himself had done in his student days. The ministry requested Marburg to name some other candidates.

Marburg responded slowly. The only other professor who might have made recommendations was the mathematician Friedrich Stegmann, who was ill. Melde asked to be excused on account of the additional burdens imposed by Stegmann's absence. Ten months went by. Finally, in March 1877, the ministry insisted on a list of nominations. Marburg now offered a list of two: Feussner and Ferdinand Braun, the Leipzig high-school teacher.

For the rest of his life Braun puzzled over why the faculty of Marburg had put an outsider on the list of nominations, especially since Melde was not particularly well disposed toward him. Since Braun had not yet qualified as a university teacher and his scientific work did not seem to him any more important than that of other potential candidates, he attributed the nomination mainly to his book *Der junge Mathematiker*, "whose exposition was so striking and clear that it revealed a born teacher," according to a Marburg student journal. No doubt the faculty also was influenced by memories of the examination for the teaching credential in 1873, which Braun had passed in outstanding fashion, particularly in physics. At that time even Melde had recognized that his student's grasp of the subject was "precise and exhaustive." In the discussions concerning the new associate professorship, Melde suddenly became "a very warm advocate of Herr Braun."[103]

In May 1877 Braun officially was offered the new position. He accepted immediately.[104] At the end of May he discussed the position in the Berlin office of the Prussian Ministry of Education with *Geheimrat* Göppert. Braun reported to his parents, "G. thought that I should be on a pleasant footing with Melde, since we were both from Fulda. I made no comment on that statement. Maybe I should have, maybe not. In any case the gentlemen seemed otherwise well informed . . . ," which is perhaps why Göppert was not surprised when Braun asked for written confirmation that he could "give his lectures in the physics institute building." In Berlin Braun also learned that his candidacy had been strongly supported by Quincke.

In late September 1877 Braun moved to Marburg. He modestly waived reimbursement for moving costs from the State of Prussia; "Since I am not married, I would consider it inappropriate to claim such reimbursement." Actually the household moves were no longer simple since his personal stock of physics apparatus had grown considerably. He had to ask Berlin for "a special room for personal scientific work in the institute." Melde's assistant was requested to give up his living quarters so that the laboratory could be provided. Franz Melde, the master of Braun's student days, now a colleague, lived on the second floor.

Marburg was half again as large as when Braun had left it only nine years before. Much of its local color had faded, but at least now Ferdinand's brother Philipp was teaching in the town and could help him get acquainted.

Under Braun's guidance, physics instruction in Marburg soon began to revive. Braun introduced free colloquia modeled after those he had seen in Berlin to supplement his new course on recent research in physics. He attracted mainly students aiming at careers in physics rather than secondary education. Roused by the competition, the lethargic Feussner reinstituted his former course on theoretical physics, while Braun, the theoretician, offered a course on topics in experimental physics.

Colleagues in other departments kept a close eye on these developments. The chemist Friedrich Fittica, for example, had been rather cool toward Braun since the publication of Braun's second Leipzig lecture on unipolarity in the *Annalen der Physik und Chemie* in 1876. Fittica was coeditor of an annual review of advances in chemistry to which Braun had referred in the final lines of his paper in no uncertain terms:

Finally, I must really protest against the viewpoint of those who would see the experiments I have described as paralleling the observations of Du Moncel, as seems to be the case in the report in the Naumann-Fittica *Jahresbericht über die Fortschritte der Chemie*. It was not my intent to place before the readers of this journal a report on phenomena that depend on how wet the substance is or perhaps on the variations in polarization that might be obtained with changing polarity when the electrodes are of unequal size.[8]

What had happened was that Du Moncel had published a paper on unipolar conduction in Paris that was obviously based on a faulty experimental setup.[336] To be compared with such a worker was particularly vexing to the meticulous Braun.

On 8 November 1878, Johann Conrad Braun died in his 80th year. This simple, modest man whose life had been devoted to his family was surrounded on his deathbed by a circle of children of which any father would be proud. The oldest son, Wunibald, recently had sold his business in Russia and moved to Frankfurt to cure his homesickness and give his children a German education. Ludwig was a successful veterinarian, Philipp soon would be confirmed in a regular teacher's position, and Adolf had just left a position as *Assessor* in the attorney general's office in Frankfurt to become one of the managing directors of the *Deutsche Hypothe-kenbank* in Meiningen. Ferdinand, the youngest son, stood at the brink of a brilliant academic career.

His scientific activity went on at the customary pace. He had made his debut before the Marburg Society for the Advancement of Natural Sciences with a reading of his Leipzig paper on unipolarity on 18 October 1877. At subsequent meetings he read other papers on the subject, demonstrating that manganese oxide also exhibited a rectifying effect and that this phenomenon did not depend on a surface layer of gas, as he was able to repeat the experiments in a vacuum with the same results. Explanations of this effect based on the assumption of electrolytic conduction also were proved erroneous. He subjected samples to unidirectional current flow for nine hours at a time but could not detect any decomposition of the crystals, which would have pointed to an electrolytic conduction mechanism. He recognized that the strong variation of contact resistance with the area of the contact could be very well exploited in microphones. "I have simply not had the opportunity to carry out the experiments before now," he wrote. (Despite the impression gained by observers on both sides of the Atlantic, his practical sense was well developed; the contrary view was a false conclusion drawn from his deliberate decision not to take out patents on any of his early work.)

Several of his Marburg papers appeared in the Society's proceedings, the *Sitzungsberichte der Gesellschaft zur Beförderung der gesamten Naturwissenschaften in Marburg*. He sent a reprint of one of them to Gustav Wiedemann in Leipzig, who had just replaced Poggendorff as editor of the *Annalen der Physik und Chemie*. Wiedemann assumed Braun was submitting the paper for publication in the *Annalen,* where it duly appeared in August 1878. Braun was horrified— he would have made his account much more detailed had he known that it was to be published in the *Annalen,*[15] but he let it stand.

Another Marburg paper, likewise read in 1878, dealt with a somewhat more theoretical subject: computing the potential of a battery from its chemical characteristics.[12] In his Würzburg work and in a paper on the unipolar properties of flames,[10] Braun had considered that the usual assumption that a battery's chemical energy was fully converted into electrical energy must be false.

Now he showed what was "in principle the correct manner of dealing with the problem[123] by proving that various electromotive energies were to be ascribed to chemical processes."[118]

This paper also appeared in the *Annalen* and "created a sensation."[132] To be sure, it was not to be compared in importance with the superficially similar finding made by Rudolf Clausius in 1850 that heat could not be converted fully into mechanical energy.[239] But Braun caused something of a commotion with an attack on another theory, which William Thomson (later Lord Kelvin) had proposed in 1851 on the basis of Helmholtz's great 1847 paper on the conservation of energy; namely, a method for computing electromotive force. Was the brash newcomer, having made kindling of the concepts of classical physics on conduction, about to shake the giant trees in the great preserve of natural science? It is recognized that the 28-year-old Braun succeeded in showing "that the W. Thomson-Helmholtz method of computing electromotive force was inadmissible."[297] He had not hesitated to take on the two giants of physics, both still living (though he did not go so far as to mention either by name), and to "break the spell by which the authority of the famous electrician, Sir William Thomson, threatened to check scientific progress in this field."[305] Other authorities stress that Braun, by his own admission primarily an experimental physicist, had in this instance made a contribution of the first order "through certain theoretical considerations"[123] that had led to "an irrefutable clarification"[121] of Thomson's theory. (Helmholtz must have thought so too. In 1882 he reworked the problem by introducing "free energy,"[337] a concept identical with Braun's "work potential" of 1878.)

Braun was also active in mathematical physics. His only two mathematical papers were completed in Marburg in 1879: "On Spherical Functions," in which he stated a new theorem about the two-body potential;[13] and "On Elliptic Oscillations," which described the motion of a particle of the ether that is part of two light rays polarized in mutually perpendicular directions.[14] (Like all physicists of the day, Braun relied on the *ether*, the all-penetrating, weightless, and completely nonviscous substance that had been postulated to account for the propagation of electromagnetic effects through space.) Mathematics, Braun said, has a charm of its own as "the transition between the limits of our senses and a more profound realm; it leads us to the very inner essence, the purely mental rule."[6] He admired the Chinese and the way in which "a very simple artifice, their 'tactile' arithmetic, had led them to decimal fractions long before the Occident," whereas occidental medieval culture earned the somewhat harsh judgment that it "had produced scarcely a single clear idea in science or mathematics" and had done "nothing to further them and little enough to sustain them."[6]

In the meantime Röntgen, who had become associate professor of theoretical physics in Strasbourg, had gone on to a full professorship in Giessen. Kundt initiated the search for a successor. His first choice was Anton Overbeck, associate professor in Halle, but Overbeck let it be known that he was not looking for a change; he either liked Halle too much or else thought he had a good chance of a promotion there.[302] The near certainty of his refusal led the ministry to reject the nomination. The search committee thereupon nominated Ferdinand Braun.

At 29, his youth played a decisive role. None but the youngest academics would face the uncertainties of a borderland university whose scientific reputation still remained to be established.[309] Kundt himself, the full professor, was only 39. Moreover, Braun's difficult position in Marburg—his preferment over Melde's own assistant Feussner—was a matter of common knowledge. Finally, a person of Braun's temperament would view Strasbourg not as a difficult post but as a challenging opportunity. He wrote Göppert in Berlin without delay:

Although no similar circumstances [to those that had led Berlin to reject Overbeck's nomination] can be said to exist in my case, nevertheless I write to ask that, so far as you are able, you should not place any obstacle to a call in my way. The reason for my taking the liberty of making this request lies in the conditions at Marburg: the probability that I should be able to conduct any experimental investigations of my own here is so small and the situation has grown so much worse that—at least in this regard— I have sometimes regretted exchanging my previous position for my present one. I do not want to bother you with details, especially since all such tales run the danger of making the teller seem petty or hateful and suspicious. Moreover, I readily admit that closer investigation would reveal a misunderstanding on one side or the other in every individual instance. That granted, I still consider my present position, no matter how I look at it, untenable *for me* in the long run. All my inclinations and scientific convictions point to experimental studies. If I am to have a chance of conducting them in a fashion that is not repugnant to me—the more so since I am forever indebted to Prof. Melde for his past intercession in my behalf—my only hope is to leave, for which a welcome opportunity now presents itself.[104]

This forthright explanation found favor in Berlin, and Braun received his invitation to Strasbourg. At the end of the 1879–1880 winter semester, he left Marburg to join the "model institute, seedbed of German physics" that August Kundt, "the unique teacher and leading experimental physicist of his day,"[206] had built up during his seven years there. In Marburg, Braun's departure gave the hapless Feussner his long-awaited chance: he was finally named associate professor. To be sure, physics instruction suffered a setback; the lectures on electricity disappeared entirely from the catalog.[159] Later on, Melde regretted that the number of mathematics and science students in Marburg had declined considerably during the 1880s.[169] During Braun's six semesters there it had grown by 50 percent.

And spread before my eyes I saw the lovely landscape in which I would bide and live for a time: the handsome city, the spacious meadows with their interlacing, splendid orchards . . .
Goethe, on arriving to study law in Strasbourg

Strasbourg, part and parcel of Europe's destiny, originally a Roman settlement, became a free city in the Holy German Empire in 1262 and property of the kings of France in 1681. Retention of certain rights and privileges made incorporation into France more or less acceptable to the German burghers—so much so that in 1815, France's weakest hour, they made no serious attempt to "return to the fatherland" or even to gain independence on the Swiss model. In 1871, *Rückgewinnung* (repossession) of Alsace became a war aim of the Germans, who could not imagine that a substantial part of the populace fancied German rule even less than French rule.

All that was new to Braun when he arrived in Strasbourg in the spring of 1880. If, like Goethe, he surveyed the countryside from the platform of the cathedral, he would have seen what appeared to be one gigantic construction site. The city's growth, the artillery damage wrought during the siege of 1870, the determination of the new authorities to put their stamp on the city's architecture all combined to make the postwar period one of Strasbourg's greatest expansion. A new railway station, a government palace, and additional university buildings went up simultaneously. But the proportion of local students went up more slowly, perhaps because the university was viewed as an agent of "Germanization." (Alsace and Lorraine were regarded as conquered territories, run by an administration headed by an imperial governor.) Even counting the sons of the new government officials, local students made up no more than 30 percent of the total. Yet no distinctions were made between local students and those from Germany proper, especially in the sciences, which remained above politics. The University of Strasbourg became the first on the Continent to cater to a student body interested chiefly in working hard.

Braun moved into a suburban apartment at No. 37 St. Urban's Way, not far from the old academy that provisionally housed the physics institute.[203] This location spared him acquaintance with the bedbugs that infested the city after the war. However, his residence between river and harbor made him prey to the other scourge of the town: mosquitoes. Goethe had tried to escape them by wearing leather leggings under his silk hose. Braun ultimately adopted the method of the tropics: a mosquito net.[320]

In the summer of 1880 Braun began his lectures on electrostatics and magnetism and on capillarity. Then followed courses on galvanism, the mechanical theory of heat, light, selected topics in physical chemistry, analytical mechanics, selected topics in theoretical physics, and thermoelectricity.[203] His working relation with Kundt went well from the first.[132] Braun was particularly impressed with Kundt's concern that some physicists relied too much on experimentation. The prevalent view among the younger physicists was that an experiment was a question directed to nature that she must answer. Kundt, the great experimentalist, warned them not to neglect theoretical analysis: it was better to carry it far enough so that a minimum of experimentation was needed to determine the numerical relations.

In the autumn of 1881 Braun moved to one of the Dacheux houses (named after their architect) on Ruprechtsauer Avenue, nearer to the university and to the new quarters of the physics institute. (All the buildings in which he lived and worked still stand.) During 1880–1882, he continued his work on the generation of electricity as an equivalent of chemical processes and on unipolarity. His work on the Thomson-Helmholtz theory, begun by experiments in Marburg, was confirmed conclusively in the better-equipped Strasbourg laboratories. Wiedemann was uncharacteristically generous with space in the *Annalen der Physik und Chemie* when Braun submitted his paper reporting these results—eight pages of tables alone.[16] But Helmholtz had anticipated him. A month earlier Helmholtz had discarded the older theory in his paper on free energy.[337] We know from Braun's subsequent writings that Braun was keenly disappointed.

It was with even greater enthusiasm that Braun's paper attacked another theory, which Franz Exner of Vienna had published in the meantime.[338] He refuted Exner's results, showing them to have been based on experimental error, by repeating Exner's own experiments under the most carefully controlled conditions. He was guided by his principle that "true relations can be established only when the observational errors to which even the most accurate measurements are subject cannot hide the regularities"—a principle he felt Exner had violated. Braun also speculated that chemical relations might be deduced from thermal properties and that Berthelot's principle in thermochemistry was but a special case of what might become a new, third law of thermodynamics. (Walther Nernst, whom similar considerations led to work that earned him the Nobel Prize in 1920, was then still in kneepants.)

Braun also continued his investigation of the rectification effect in semiconductors and expanded it to include alternating and interrupted direct currents in addition to the pure direct currents he had used in his earlier work. He was goaded into this work by the incomplete paper that Wiedemann inadvertently had reproduced in the *Annalen,* which had led a Swiss doctoral candidate in Göttingen, Hans Meyer, to announce that he was unable to reproduce Braun's anomalies when direct and alternating currents were applied simultaneously. At the same time Braun had to deal with a more serious criticism—from Gustav Wiedemann himself.

In a revised edition of his *Theory of Electricity*, Wiedemann had thrown some doubts on Braun's results, citing possible sources of experimental error that Braun's meticulous observations already had eliminated. Braun's measured and firm reply, which duly appeared in Wiedemann's *Annalen,* reiterated that neither local heating, surface changes, nor other artifacts could account for what was without doubt a real phenomenon.[17] He underlined his contention with "a series of new numerical observations made recently before fellow scientists here, Prof. Kundt among them."

"Some Remarks on Unipolar Conduction in Solids," published in the *Annalen* in January 1883, was Braun's fourth and last publication on the rectifier effect.[17] All that could be done experimentally and theoretically with the means at his disposal, he had done—and done so well that little more was to be added for half a century. His last observation improved on his earlier finding that the effect occurred even when the current lasted only 1/500 second. He sent the current through a metal pendulum, so that the passage of its tip through a mercury trough created an electrical contact whose duration could be computed with great exactitude— a simple and ingenious way of measuring minute intervals of time.

The final sentence in Braun's paper on unipolarity states, "At present I do not dare to propose an explanation for these observations, any more than I did on previous occasions; I only want to emphasize again and again that they are not to be explained by previously known facts."[17] This statement calls to mind Newton's words about gravity: "I have so far not been able to deduce the cause of this phenomenon from the observations. And I frame no hypotheses."[339]

At the end of the 1882–1883 winter semester a full professorship fell open at the Technical University of Karlsruhe, capital of the Grand Duchy of Baden, when the incumbent, Leonhard Sohncke, accepted a position at the University of Jena. The search committee considered three candidates: Georg Recknagel, principal of a polytechnic school in Kaiserslautern; Ferdinand Braun, associate professor in Strasbourg; and Anton Overbeck, associate professor in Halle. Recknagel was the oldest and could be expected to remain longer than an ambitious younger man who might be tempted by a subsequent call from a regular university. Moreover, he had some experience in meteorology, a subject that traditionally was associated with the chair of physics at Karlsruhe. But the authorities pointed out that only forestry students were still expected to attend lectures on meteorology and that a separate Central Bureau for Meteorology and Hydrography was about to be instituted in the Grand Duchy, so that, in the future, the physics professor need not be a meteorologist. The other two candidates, young Braun and young Overbeck, would not only be satisfied with lower salaries, but had both specialized in electricity. Other technical universities—in Berlin-Charlottenburg, Stuttgart, Darmstadt —were instituting courses and starting departments of electrical engineering. Karlsruhe had no such plans and therefore was under pressure to have a physics department whose head had a good knowledge of the subject. The search committee's vote was split evenly: six for Recknagel and six for Braun. The ministry chose Braun.

After five and a half years as associate professor, Braun, not yet 33, would receive his first offer of a full professorship. Was it coincidence or a good omen that a member of the search committee also had been a pupil of the remarkable Dr. Wilheim Gies? He was Wilheim Schell, who, a quarter of a century before, had helped another pupil of Gies, Franz Melde, to the professorship at Marburg. Another member was Adolph Knop, the mineralogist to whom Gies had sent the 15-year-old Ferdinand Braun's manuscript of a textbook on crystallography for evaluation. (All three—Braun, Schell, and Melde—were to be reunited in a memorial volume that was prepared for their beloved teacher a few months later on the occasion of Gies's 70th birthday.[146]) There was the usual warm recommendation from Quincke. Still another member of the search committee, Dr. Carl Engler, was destined to play a role in Braun's life.

In the same year, Braun had had the satisfaction of repaying part of his debt to his oldest brother Wunibald by helping him to make a most promising business connection. He had learned that a Würzburg firm specializing in scientific instruments, owned by Eugen Hartmann of the Swabian town of Nürtingen, was not operating profitably even though it was doing a lively amount of business. Revenue was being dissipated by costly experiments and the obligation to make innovations. The company had to mannfacture some staple product to generate needed capital. Such a device was in fact available—a recording galvanometer that Kohlrausch, Quincke's successor in Würzburg, had offered Hartmann. With the rise of the electrical industry, the invention stood a good chance of becoming a commercial success. Braun drew the opportunity to his brother's attention. Wunibald became a partner—a silent one at first—in a new firm, E. Hartmann & Co.[236] That was the predecessor of one of the leading enterprises in the field of measurement and standardization, today's Hartmann & Braun AG of Frankfurt.

Ferdinand Braun had been a customer of Eugen Hartmann's before his brother joined the firm. In the memoirs of Hartmann's first apprentice is a record that in 1882 the firm made for Braun what was probably the first German ac transformer with a closed iron core. The core cross section was a respectable 60 square centimeters. There was no ac generator in Würzburg, and the transformer had to be sent to Schuckert in Berlin for testing. Why Braun ordered it and what became of it we do not know.

6

CREATIVE YEARS: PROFESSOR IN KARLSRUHE, TÜBINGEN, AND STRASBOURG (1883–1918)

The rail journey from Strasbourg to Karlsruhe is short, but the train, rushing through the flat country between the Black Forest and the Vosges Mountains, carried Ferdinand Braun to a new world. After nearly six years in someone else's department, he would be his own master at last. In a society in which the competition for titles sometimes seemed to be an end in itself—however indifferent Braun might have been to it personally—he could not disregard the fact that he had become a full professor. Not only had his monthly salary increased from 240 to 416 marks overnight, but he had also gone from the newest and least established physics institute in Germany to one of the oldest and best established. The Technical University of Karlsruhe was the third such institution in Central Europe and the first in Germany to have a physics laboratory equipped from public funds. Famous scholars had preceded Braun, among them Gustav Wiedemann, who had gone to Leipzig from Karlsruhe in 1870.

To be sure, Braun's new domain could not compare with similar departments at regular universities. There were never more than 300 students at Karlsruhe altogether, and never more than 9 of them specialized in mathematics and the natural sciences, a division that was subdivided into no fewer than seven departments. But experimental physics was a required subject for students in other departments (engineering, mechanics, structures, chemistry, and forestry) and the lectures were well attended.

At 11 A.M. on Monday through Thursday Braun lectured on experimental physics, and on Wednesdays at 5 P.M. on meteorology, in a large lecture hall that seated 200. Laboratory exercises for physics students took place in the basement laboratory from 2 to 5 P.M. on Mondays and Thursdays. In charge of the laboratories was Dr. August Schleiermacher, Braun's old acquaintance from his first year at Strasbourg, an assistant whom he had taken over from his predecessor Sohncke. (Schleiermacher, who came from Darmstadt, was to spend his long career in Karlsruhe: he had gone there in 1881, became a professor in 1894, and was still living there in 1950, when the institution celebrated its 125th anniversary.)

Appointment to a full professorship called first of all for an inaugural lecture. On the afternoon of 23 April 1883 Braun spoke on the question, Should departments of electrical engineering be instituted at technical universities, along with the existing departments of machine design and engineering? Electricity was traditionally taught as part of physics, from a "scholarly" viewpoint. Its practical application was left to instructors who, as Braun put it, "know the rules for calculating current distribution in a given network, how much current is needed to light a bulb, what the conventional machines are—who in short possess a great deal of practical and useful information," but who scarcely could be called electrical engineers.

Yet as more and more arc lights went up in city squares, factories, and railway stations, "bringing electrical engineering into great public favor and placing great hopes in it," it became increasingly obvious that anyone who undertook to exploit this new source of energy depended "more than any other professional on the kind of theoretical knowledge that integrated hard-won data and extensive observations most concisely: knowledge that served to guard him against waste of time and against errors."[18]

The need for departments of electrical engineering had been endorsed by Werner Siemens, and in fact some of the technical universities had started such departments, but not all were anxious to explore new territory. In his inaugural lecture Braun cautioned that "electrical engineering is not ready for such treatment under present conditions, when it is still developing and all that is new derives directly from theory, and when even the underlying physical principles can scarcely be said to have been reduced to a few axioms from which everything else could be deduced." Scientific education at a technical university should not be limited merely to fulfilling technology's current needs, "but if possible should be such as to last the present engineering students for the rest of their careers." Since that was then an unattainable goal, Braun thought it would be better to continue to teach electricity as part of physics.

Braun knew that this viewpoint was shared by several outstanding professors of the time, the academic senate at Karlsruhe, and the government of Baden, which would have had to find the money for any new department of electrical engineering. The authorities also favored continuing electrical engineering as a part of physics; after all, they had picked Braun precisely because of his practical knowledge of electrical engineering. This argument was made in an official document of which Braun was unaware, or else he well might have pressed more vigorously for the 50 marks more a month that he had originally demanded.[305]

Under these circumstances, the government of Baden could well afford to provide what Braun in all innocence considered "a most generous and accommodating" subsidy of 2,000 marks to establish his special course in electrical engineering, which would supplement a course of "applied electricity" already being offered at the Karlsruhe *Gewerbehalle* ("Industrial Academy") by Dr. Meidinger.[209] This arrangement, which could serve as a transition toward a possible future separation between physics and electrical engineering at the Technical University of Karlsruhe, called for lectures and for "the equipping of a laboratory similar to a physics laboratory that would give the students an opportunity to obtain practical experience in the application of the methods they had learned and to test their theoretical knowledge."[305]

To provide the space for the new laboratory, the department of mathematics kindly offered part of its lecture room, which was too large for its own classes.[213] A wooden partition was put up. From then on the mathematicians had the dubious pleasure of lecturing over the crackling of electrical sparks and the hum of induction machines.

A permanent part of the laboratory was a dynamo made by Siemens & Halske, which the college had acquired in an unusual way. Braun's predecessor Sohncke had once happened to check an old inventory and found that an old map was missing. He then went systematically through all the old inventories looking for missing items. He found that someone had borrowed a valuable machine tool twenty years before and never returned it. The hapless borrower was found and paid 1,000 marks in compensation, an amount that was just sufficient to buy the dynamo that Sohncke repeatedly and unsuccessfully had requested from the government.[213]

Braun's request for 2,000 marks for his special course had been endorsed by the administration because the unprecedented number of nine students had enrolled in the physics laboratory course.[305] It was in these intimate surroundings that Braun created his first "school" in 1883 and 1884, a school whose influence was to be felt into the twentieth century. Besides the 33-year-old "master" Prof. Braun and his 26-year-old assistant Dr. Schleiermacher, there were 28-year-old Dr. Carl Feussner, later the resistive-materials expert of Germany's National Physics Laboratories; 27-year-old Albert Gockel, later *Privatdozent* in Freiburg in Switzerland and professor of physics and meteorology at the University of Padua; and 25-year-old Otto Ehrhardt.[156] The department's activity centered on high-temperature physics and electricity, fields with which Braun was well acquainted. He made good use both of his electrical experience and of his earlier work on alloys. The choice of these topics showed that he knew exactly what was expected of a science department at a technical university: active participation in the development of electricity and recognition of the needs of industry, which just then meant particular emphasis on metallurgy, especially smelting. The contributions of the Braun school are exemplified by several papers: "Thermal Electricity of Molten Metals" (Braun); "On the Dependence of Heat Radiation on Temperature and Stephan's Law" (Schleiermacher); "On the Relations of Peltier Heat and the Efficiency of Galvanic Elements" (doctoral dissertation of Gockel); and "On the Relations of Specific Heat and the Heat of Melting at High Temperatures" (doctoral dissertation of Ehrhardt). In addition, Braun designed and developed a galvanometer and an electrical pyrometer; Feussner built a voltmeter; Ehrhardt, a calorimeter; and Schleiermacher, a thermostat. Braun's school obviously was contributing to the development of technology.

The group's scientific orientation was toward engineering research; that is, each investigation had an eye to practical applications from the start. That is not to say that Braun neglected the fundamental research that prevailed at the regular universities, where the philosophy was, as he put it, "an inclination to penetrate the secrets of nature, to be led from one question to another without—fortunately for the progress of science—giving a thought to any possible practical significance."[6] To begin with, all faculty members at the Technical University of Karlsruhe were themselves graduates of universities. Even the graduate students Gockel and Ehrhardt were preparing for doctoral examinations at universities, one at Heidelberg and the other at Giessen.

Relations between Braun and the members of his departmeent must have been most cordial. In a paper his assistant, Dr. Schleiermacher, acknowledged Braun's help and support in terms normally associated only with dissertations.[244] It must have given Braun particular satisfaction to have as a colleague Dr. Carl Feussner, a brother of the same Dr. Wilhelm Feussner with whom Braun had had such precarious relations in Marburg.[156] And it was a student from Karlsruhe, Otto Ehrhardt, in whose doctoral dissertation Braun's help was first acknowledged.

In September 1883 Braun attended the International Electrical Exhibition in Vienna and published a report on it in several installments in the *Centralzeitung für Optik und Mechanik,* to which he was a regular contributor on electrical subjects.[19] "The Siemens electrified railroad definitely aroused the greatest public interest," he wrote. He also described what was presumably the first electric traffic light: "On Artillery Avenue, where the traffic is thickest, there is an electrical signal consisting of a glass disk, mounted on a pole, which serves to halt the carriages. At night a lantern is lit behind the disk, which shows a red light as a stop signal but normally sends forth a green light." Braun was particularly enthusiastic about the modern Weston lamps. "Their light spreads all over the imperial city; even from top of the Leopoldsberg, two hours away, one can see their reflection on the low clouds on a rainy day." A special effect by which electric light was sent through an artificial waterfall left him unimpressed, but he noted "with great astonishment the immense magnification ($\times 10,000$ linearly and $\times 100,000,000$ in area) attainable in projecting images from a microscope by means of the electric light."

From Braun's meticulous description we get a glimpse of what were considered the wonders of electrical engineering to come, as exemplified by the exhibits: electrical illumination in private apartments, a picture gallery lit by electricity, a device to turn the heat of an ordinary oven into electricity by thermoelectric means, and safety measures such as an electrically actuated water-sprinkler system for theaters.

Of central importance to the high-temperature physics carried on at Karlsruhe under Braun was the so-called muffle furnace. It had been built by Ehrhardt and the "able and loyal"[213] mechanic of the institute, Gottlieb Martin. The oven comprised a system of heating chambers, one within the next, heated by gas or coal; the innermost, shielded by the muffles against the direct action of the heating fuel, could be heated to temperatures up to 1,100°C.[243] The great exhaust funnel and the rugged iron under-carriage gave the room a factory-like appearance. The undercarriage served to transport the molten materials as quickly as possible, before they cooled, to the research apparatus. After some practice, the Karlsruhe team was able to whisk the clay container of the molten materials into the undercarriage below in less than two seconds.[243]

The first use for the new equipment was to determine the highest temperatures that could be reached in the oven. From the porcelain factories of Meissen and Berlin, the laboratory borrowed two "atmospheric thermometers," then costly rarities. Braun found them very fragile and unwieldy. He saw a need for rugged, reliable, and handy apparatus that could stand up to the rough treatment it would get in an industrial furnace. He contrived one, consisting essentially of a piece of platinum wire whose resistance change with temperature was measured by means of a galvanometer.[30] The principle was not new; the Braun electrical pyrometer was not an invention. But a great deal of inventive genius went into its construction. Most rugged galvanometers then available drew such a heavy current that the temperature of the platinum wire rose under the action of the meter current itself, which gave wrong readings. Braun's device overcame this difficulty by using currents so feeble that the effects of terrestrial magnetism had to be taken into account. To achieve this goal, he had to redesign the galvanometer.

Calibration of the pyrometer was difficult. In the range from 400 to 1,100°C the calibration took fifteen hours at the muffle oven. Most of this work was done by Dr. Schleiermacher. The finished pyrometer consisted of the platinum wire in a fireproof container that could be installed anywhere, even inside an industrial furnace. The galvanometer could be some distance away, perhaps at a central switchboard. A battery provided the measurement current. This arrangement, the Braun electrical pyrometer, was thus an early instance of telemetry, which is an essential component in today's remote-control systems (figure 8).

The pyrometer passed its ordeal by fire (in a literal sense) in the course of research done by the Braun school at Karlsruhe during 1883 and 1884. Ehrhardt used it to determine the previously unknown melting points of three salts, a continuation of the work on salt solutions that Braun had done in Würzburg. Gockel's dissertation derived from another earlier work by Braun, on the generation of electricity as an equivalent of chemical processes.[12] Schleiermacher used the Braun pyrometer in his work on heat radiation, and Braun used it in the work for his paper, "On the Thermoelectricity of Molten Metals."[20] The object was to explain why a voltage appears across the ends of two pieces of dissimilar metals when the other two ends are joined together and the junction is heated. Here again, as he had done with the salts, Braun used metals in the liquid state, because "structural relations are simpler" for liquids than for solids.[3]

"On the Thermoelectricity of Molten Metals" was Braun's only scientific paper from the Karlsruhe years. It represented the Braun school at its highest level of development, since it was both applied and fundamental and took the exigencies of industry into account, for instance by pointing out that the currents that arise in liquefied metals might be useful in the measurement of very high temperatures.[20] The combination of theory and practice is seen in the following passage: "There is every indication that thermoelectrical generation is an intermolecular process that depends on atomic number and possibly on the motion of atoms within a molecule. One is led to the thought that all gross mechanical transformations, which likewise exercise a great influence on the thermoelectrical behavior of a solid and manifest themselves in strain, bending, hardness, and so forth, are related to intermolecular—which is to say, chemical—processes."[20]

Elektrische Pyrometer nach Prof. Braun.

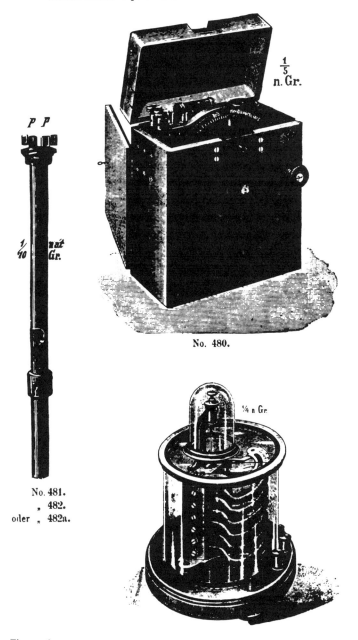

$\frac{1}{5}$ n. Gr.

No. 480.

¼ n Gr.

No. 481.
„ 482.
oder „ 482a.

Figure 8
Page from Hartmann & Braun 1893 catalog illustrating Braun's pyrometer.

The experimental part of the work was also ingenious. Because of the great range of temperatures that had to be explored, Braun used graphite (a material that could withstand the highest temperatures) as a probe to generate the measurement current, which was then carried by platinum and lead. But once again the results fell short of his expectations: he was unable to deduce a law from his experiments. "When it comes to thermoelectricity, we are probably further from a real insight into the basic principles than I (and most physicists with me) would have thought."[20] Braun considered his findings so significant that he submitted the results to the highest scientific authority in Germany: the Royal Prussian Academy of Sciences. The paper was read at the meeting of 12 March 1885 by its president, Hermann von Helmholtz, who had just been raised to the nobility.

This achievement moved Braun into the small circle of recognized scientists, just as the appearance of his doctoral dissertation in the *Annalen* had first moved him into the group of serious scientific investigators. No less important was the personal side of the success. The paper was an appropriate celebration of his return to the university as well as an introduction to his new place of work. It was also a fine wedding present for the young woman he was to marry ten days later.

The marriage had come about in the following way. Braun had become friendly with the professor of chemical engineering at Karlsruhe, Dr. Carl Engler, a jolly and friendly man, always ready for a joke. In a spirit of raillery Engler prophesied that Braun would soon marry and lightly suggested that he had already picked a wife for him. If he could manage that, said Braun, he could serve as godfather to his firstborn.[321] This jovial exchange took place on 11 July 1884, at a party given by the Polytechnic Union of Karlsruhe to honor Engler, who had served as director of the Technical University during the previous year. The minutes record that both men made "humorous speeches in the course of the evening, alternating with jolly student songs."[216]

A young bachelor's social life was unbelievably circumscribed in the 1880s. Social contacts, especially among young men and women, were possible only in large groups, properly chaperoned by reliable married couples.[135] The etiquette of the day even required that on a train journey toward the starting point of a country ramble the young men and women had to sit in separate compartments. Opportunities for meeting members of the opposite sex were largely limited to heavily chaperoned balls in the winter and to garden parties in the summer. The odds were thus against Engler in his endeavor to find a bride for Braun, a hardened bachelor of 34. But Engler, a skilled experimentalist, was a practical man. He started by taking Braun to a garden party at Lahr, at the country house of his father-in-law, Adolph Friedrich Bader.

Bader, scion of an old Swabian family of officers and civil servants, had started the first cigar factory in that part of the world in 1840, just as the fashion swung away from the taking of snuff and the smoking of pipes.[217] He became quite wealthy. In addition to his villa in Lahr, he had a summer residence on an estate nearby, where his large family and his many acquaintances assembled each Sunday to enjoy themselves. It was here that Braun met his future wife, Amalie Bühler, who was in fact a niece of Engler's wife, the daughter of the cigar king.[323] Amalie's father, Julius Albert Bühler, the husband of the cigar king's other daughter (also named Amalie), was a partner in his father-in-law's business. Before meeting her future husband, Amalie Bühler had led the uneventful life of a daughter of a "better" family. Her sister Martha married a Wilhelmshaven physician, Dr. Adolf Schmidtmann, who later became the head of the health department of the Prussian Ministry of Education. Amalie's brother Conrad Bühler became a banker in New York.

Ferdinand Braun made up his mind to acquire a small competence by exploiting his pyrometer commercially. It was not an immediate success. Development of a production model dragged along very slowly.[30] Braun had turned the project over to his brother Wunibald as a sort of second installment of the "personal repayment" of his debt. In 1884 Wunibald had become an active partner in E. Hartmann & Co. The company's main product, the recording galvanometer designed by Prof. Kohlrausch, had become a great success, and Wunibald had taken over as business manager. The relocation of the growing firm—now called Hartmann & Braun—from Würzburg to the Frankfurt suburb of Bockenheim further delayed the development of the pyrometer. When work on it was resumed, the engineer in charge of the project found he needed several months to design and manufacture the requisite precision components.[30] A selection of these pyrometers finally was exhibited at the Frankfurt International Electrical Exposition of 1891. They were competitive products capable of satisfying a demand for a versatile line of high-precision instruments.

Braun came to appreciate the beauty of Karlsruhe during his two years there. The elegant palace that the grand dukes of Baden had laid out some 180 years before, the broad avenues, and the many gardens and parks all called forth a sympathetic response. Across the way from Braun's apartment stood the imposing façades of two public buildings.[209] Karlsruhe was then a town of 57,000 inhabitants. It had a steam railroad, and a new system of "hygienic sewers" bore witness to the city's progressive attitude. A lively debate raged in the newspapers over whether the newfangled American telephone should be introduced.

The small capital city was permeated by an atmosphere of progress and a feeling that an era of stagnation had come to an end. As Braun put it,

The age of discovery and geographical exploration was accompanied by another sort of revolution, as if the finding that things were quite different from what had been believed through the ages with dogmatic certainty in a narrow-minded and shortsighted way had given the human spirit courage to break with long-established views in other directions as well. Our outlook grew wider and more liberated once science dared to break free of the fetters of scholastic syllogisms. We no longer paid heed to the dire forecasts of how we should go wrong and get lost in the labyrinth of an imaginary world once we had left the narrow path of proven truth, with no faith nor any other restraint. . . . Science's struggle for independence was rewarded by far more success than our guardians—religious dogma and pedantic scholasticity—found convenient.[6]

Of Braun's salary, 50 marks represented remuneration for his work as meteorologist. It was not an onerous task, and he found the weekly meteorological lecture for forestry students instructive for the teacher also. But it took him months to rid himself of the requirement of making weather forecasts. Braun's conception of his duties as scientific adviser to the new Central Office for Meteorology and Hydrography of the Grand Duchy of Baden was quite different from the government's. The government expected him to contribute information that would aid in the daily forecast "in cases of difficulty," a thankless and time-consuming job that could have easily damaged his reputation if the officials chose to make their scientific adviser the scapegoat for inaccurate predictions.

Not until Braun had resolutely pointed out the potentially harmful effects of that sort of collaboration on his proper duties as professor of physics did the ministry agree to exempt him from direct participation in the work of the weather bureau. In place of a Central Meteorological Bureau he offered to set up a Central Bureau for Electricity at the physics institute and to become its director. The offer was accepted but proved to be ahead of its time. Throughout the remainder of Braun's stay at Karlsruhe none of the Grand Duke's subjects found it necessary to consult the new bureau.[305]

The misgivings of some members of the appointment committee as to how long a young candidate would stay at Karlsruhe proved to be well founded. A chair of physics at the University of Tübingen became vacant at the end of the summer semester in 1884 with the retirement of Prof. Eduard Reusch. Leading the list of candidates was Prof. August Töpler, 46, of the Technical University of Dresden. He was an experienced man who had built up or completely reorganized a physics institute on three previous occasions, a task that awaited the new appointee at Tübingen as well.[306] The prospect of doing it for the fourth time did not appeal to Töpler, and he declined the offer. The second candidate was Anton Overbeck, associate professor at the University of Halle, who had been in competition with Braun twice before. The Tübingen faculty knew that Overbeck had steadily refused associate professorships elsewhere, but thought he would not reject a full professorship. They were wrong; he refused again.

Ferdinand Braun was the last choice because the faculty assumed that he would not come to Tübingen unless he could be assured of a new physics institute to match the one he had at his disposal at Karlsruhe; at Tübingen such an organization was only in the proposal stage and might remain so for years. On the basis of practical ability and experience, Braun was considered the equal of Overbeck, who was four years older, and he was known to be an excellent lecturer besides.[306] The recommendation particularly singled out Braun's work on the generation of electricity as the equivalent of chemical processes, in which he had not hesitated to dispute the authority of Sir William Thomson, and found special praise for his book on mathematics and science for young people. Braun scarcely could have expected that his little book would aid him in the attainment of three professorships.

The negotiations were short. Although Braun knew full well how badly the old physics institute at Tübingen was housed and how much his own work would suffer during the construction of a new building, he accepted the offer. The chance to get back to a university overshadowed all other considerations.

His monthly salary was set at 583 marks, and he was to have living quarters in the institute. The government in Stuttgart, the capital of Württemberg, promised to do what it could to accelerate the plans for a new building and provide funds for the acquisition of new equipment. Braun would not start until 1 April 1885, so that a replacement could be found at Karlsruhe. With Braun's advice, the Karlsruhe faculty proposed three candidates: Overbeck; *Privatdozent* Heinrich Hertz in Kiel; and associate professor F. Himstedt in Freiburg. Overbeck refused once again and Hertz received the appointment. It proved to be an excellent bargain for Karlsruhe. The 27-year-old physicist, the son of well-to-do parents, was miserable in Kiel. He accepted a monthly salary of 233 marks, as compared to Braun's 416 marks. He could be expected to continue to emphasize electricity, since that was his own specialty, and he soon fulfilled the "great promise" of which his teacher Hermann von Helmholtz had spoken in recommending him: he carried out what was perhaps the most famous piece of research ever undertaken at Karlsruhe. In the very lecture hall that Braun had carved out of the space so generously provided by the mathematics department, Hertz three years later demonstrated the existence of electromagnetic waves.

Ferdinand Braun and Amalie Bühler were engaged during the Christmas vacation of 1884 and celebrated their marriage on 23 May 1885 in Lahr. Braun presented his wife with a castle as a wedding present. In a place that could not have been more beautiful, high above the rooftops of Tübingen, the couple set up housekeeping. Many of the rooms were dark and full of odd angles, but the location was superb. The site had been the domain of the counts of Tübingen since the eleventh century, and through the battlements the beautiful city and the sparkling river Neckar could be seen below.

Quite disturbing was the realization that the physics institute adjacent to the official residence was in bad shape. The lecture hall, two floors up in the cold northeast tower of the castle, was circular and humid. The two workrooms adjacent to it were so crammed with apparatus that some of it overflowed into a busy corridor that led to the university library.[305] But what made the Tübingen Physics Institute unique among German institutions of higher education was that none of its rooms could be heated. The most important scientific subjects could be taught only during the summer, which was the shorter and less well attended semester, so that some important topics in physics could be reviewed only briefly or had to be left out altogether. During the winter semester, Experimental Physics with Laboratory Exercises was merely an entry in the catalog. "Even at the beginning of the summer semester, when the days were still cold, it was necessary to use a small stove to warm the professor at least." The stove was put next to the demonstration table and was vented through a nearby window.[305]

"This morning I saw the institute at the polytechnic and was very pleased with it," wrote Heinrich Hertz in his diary for 30 March 1885, after his first visit to the rooms that had been vacated by Ferdinand Braun at Karlsruhe.[246] Meanwhile, Braun in Tübingen must have felt like someone who had exchanged a palace for a mud hut. Although his predecessor had given some thought to heating at least the physics institute proper, there was no chance of carrying out the more costly project of heating the rooms in the tower, whose top floor housed the astronomical observatory. The solution was to utilize a room in the official residence next door, which had been turned over to the institute. This makeshift laboratory was used for both large-scale and the most delicate experiments, an arrangement that led to bitter feuds among its users. In winter the room also served as a small lecture hall, with chairs carried in and out for each lecture.[305] Small wonder that Reusch, Braun's predecessor, undertook nothing that required additional space. Demonstration experiments requiring a large amount of space had not been done in Tübingen for years. Work for doctoral dissertations was undertaken only exceptionally, which meant that the institute director labored under self-imposed restrictions—an intolerable state of affairs.

Braun would not put up with it. In September 1885 he took a town official, government architect Berner, on an eighteen-day trip around all the physics institutes that had gone up in recent years—in Graz, Vienna, Prague, Dresden, Leipzig, Berlin, and Hanover. Travel expenses were paid by the government because "a mistake now could lead to expensive later modifications and be disastrous for generations to come."[305] The ideas that Braun and Berner gathered on this trip led to a new plan that has stood the Tübingen Physics Institute in good stead to the present day.

Once again Braun had to give an inaugural lecture, on "Law, Theory, and Hypothesis in Physics," in which he delineated the three stages of scientific knowledge. "The highest stage, law, demands simplicity. Where this is not the case, the investigator is probably still laboring in the realm of qualitative research." The next step, theory, is better the more firmly it is based on observations that can be interpreted unequivocally. Such a theory "gratifies by the symmetry of its perfection and its 'practical' (in the least pejorative sense) utility," for instance, when various instruments constructed according to different principles all yield the same result for a given observation. Hypothesis serves as the connecting link between law and theory: "It facilitates representation, it opens perspectives, it gives life to dry facts much as elves and nymphs render lifeless nature more pleasing to the imagination." To be sure, one must know the dangers of leaving the sure basis of quantitative experimentation or strict theory. Yet hypothesis is indispensable; it resembles

| dem befruchtenden Regen, der im Schmutze selbst zu Schmutz wird, doch auf gutem Acker Segen bringt und jederman zu Nutz wird. | The fructifying rain Turned by dirt to its own kind, Yet showers blessings on the grain, Benefiting all mankind. |

And so, with excursions into philosophy and theology, Braun brought his lecture to a close by a somewhat abrupt reference to the latter subject: "A law of nature is something quite different from the 'principles' of philosophy. It is an expression that encompasses a substantial number of *facts*."[21]

The new professor's introduction to Tübingen's social life came when he was admitted to the Tuesday Club, a weekly meeting of almost all the professors and their assistants at the *Lamb* near the town market[223] (figure 9). Its main purpose was "interdisciplinary" knowledge. Each member, in alphabetical order, had to give a lecture, a procedure that produced a somewhat motley series of topics: "On Heat Stroke" was followed by "Hearing Damage among Railroad Workers"; "Obesity" came after "Factory Work on Sundays."

Braun's turn did not come for nearly two years, until the end of 1888; his subject, "What Do We Call Large, What Small?"[223] The lecture has not been preserved. Doubtless it was related to an earlier lecture given before a more professionally oriented seminar of scientific colleagues in Tübingen, the Medical and Scientific Club, on "Limits of Microscopic Viewing." In that lecture Braun had carried the concept of smallness to its physical limits. Once the wave nature of light had been established, it was clear that no object could be made visible that was smaller than the wavelength of the light used to view it. The problem of overcoming that limitation occupied Braun as long as he lived. Later he actually came close to solving it.

At first, the new professor's scientific endeavors had to be modest. The old building was ill suited to experimentation, and construction of the new institute took up most of his time until 1889. So he theorized instead, for instance, about the reversibility of solution processes.[22] Increasing the pressure in some saturated solutions, for example, ammonium chloride, would lead to precipitation of crystals; but in sodium chloride precipitation would be decreased by an increase in pressure, and the solution would be capable of accepting more salt. Braun chafed under the limitations that prevented him from testing this theoretical result. Even if there had been space, it would have been difficult to maintain the requisite constant ambient temperature in laboratory rooms that were subject to every whim of the weather. Not until the coming of the frost and snow in the winter of 1885–1886 could something like an extended period of constant temperature be achieved. Then

Figure 9
Ferdinand Braun, age 36, from an album of the Tuesday Society of
Tübingen.

Braun filled a wooden bucket with snow.[22] As the snow melted, new snow was added, so that a temperature of 0°C was maintained.

A paper based on these studies, "Investigations of the Solubility of Solids and the Related Changes in Volume and Energy," was submitted to the Royal Bavarian Academy of Sciences in Munich on 5 July 1886. It must have been the strangest investigation to be undertaken at any German university that winter. "The room's temperature was unusually favorable," reported Braun blandly. "It remained steady at −1°C sometimes for as long as three weeks in a row." The humidity of the Tübingen castle tower likewise favored the investigation: "Salt solutions left in the open sometimes do not precipitate crystals for weeks; filter paper saturated with salt solutions remains damp for days."[22]

The end of winter brought an end to this "stupid experimentation."[104] The wooden bucket was set aside. Another series of observations was made in the spring at 15°C; and the experiments were continued the following winter, again at −1°C. Finally, on 8 April 1887, Braun was able to send a paper to the *Annalen der Physik und Chemie* on his "pioneering formulation of the dependence of solubility on pressure,"[123] which he had deduced theoretically and now submitted under the title, "On the Decrease of Compressibility of a Solution of Ammonium Chloride with Increasing Temperature."[23]

Further thought about these relations led Braun to a theoretical proposition, which he published in the summer of 1887 in Ostwald's *Zeitschrift für physikalische Chemie*.[25] He maintained that a system in equilibrium (say, a saturated salt solution) would always yield to an external constraint such as pressure in the direction of "least resistance"; for instance, the solution would precipitate salt or would accept additional salt according to which process the system's constitution made easier.

This "general qualitative law of state changes," which Braun's paper illustrated by several examples, was a significant scientific achievement, but, unbeknownst to him, this law had been announced by a French chemist, H. L. Le Châtelier, some three years earlier.[340] Braun later acknowledged that Le Châtelier's 1884 formulation and his own had been identical almost word for word.[237] However, Le Châtelier had limited the law to systems in chemically stable equilibrium, whereas Braun wished to extend it to a number of systems, including electrical ones, and so he saw no reason not to let his contribution stand and even be reprinted in various subsequent publications.

The scientific community seemed to agree with Braun's extension of the principle, which appeared in some texts as the Le Châtelier-Braun principle. Having a discovery known by one's name is a source of considerable satisfaction to any researcher. It turned out after some time, however, that the principle was fully applicable only to the systems studied by Le Châtelier, and Braun's name is no longer associated with the principle.

A curious computational error came to light during this period as Braun was preparing the second edition of *Der junge Mathematiker und Naturforscher*. In calculating the date of the Easter term for the book, Braun had not noticed that 1870, 1874, 1878, and so on were not divisible by four and hence were not leap years, so that the results were wrong. Such computational details plagued Braun all his life. Even posthumously the headline of a 1939 *New York Sun* article about him was "Wizard Hated Mathematics."[129] He had to take the ragging of some of his mathematician colleagues, who maintained that the only time his computations came out right was when he had made two errors that cancelled each other. They made up stories about him. Jonathan Zenneck, Braun's student and later assistant, told the whimsical tale that Braun, when faced with the multiplication 2 × 25 during a lecture, rounded it off to 2 × 30, wrote 60 as the answer, then added, "But since we took 2 × 30 before instead of 2 × 25, the actual result will be nearer to 50."[135]

This quirk was not, of course, due to lack of intellect but to the intensity of his concentration on the physical problem at hand; exact numerical values became secondary. This attitude carried over into his experimental procedure, in which his objective was to establish the physical relations rather than the highest possible precision in each measurement.[324] For instance, Braun announced an order-of-magnitude figure for the compressibility of rock salt as 0.000005. The more exacting Röntgen in Giessen, the acknowledged expert on compressibility, pointed out in the *Annalen der Physik und Chemie* that the correct number was 0.0000015. Braun admitted his "error" but added that he was not particularly concerned with the exact value, and that even an error of 100 percent would make a difference of only 1 percent in the result.[31] But the episode must have been embarrassing to Braun. On its heels came another mistake, when the mathematician Jakob Lüroth found an algebraic error in the same paper, so that Braun had to publish yet another correction.[34] "Dear God, if only I had your mathematical and theoretical knowledge as well as the interest and the persistence to go with it!", Braun had written ruefully on 10 April 1887, to his friend Leo Graetz.[104] (In a speech Braun made some ten years later he remarked about Faraday, "He lacked mathematical training. Perhaps that is precisely why he grasped the facts in the most simple, direct manner . . . ."[67])

While the conditions in Tübingen were forcing Braun to perform experiments in a wooden bucket filled with melting snow, Hertz in Karlsruhe achieved his epoch-making discovery of electromagnetic-wave propagation, experiments that Braun later characterized as ingenious and penetrating and as an "immortal achievement."[68] Hertz's paper appeared in the same issue of the *Annalen der Physik und Chemie* in which Braun reported the results with his wooden bucket.[248] Under these circumstances it is perhaps understandable why Braun, who was later to play such an important part in the practical application of Hertzian waves, seemed so indifferent to them at first.

On 4 April 1888, Hertz visited Tübingen on a walking tour that was interrupted by a late snowstorm. Before returning to Karlsruhe by train, Hertz decided to visit Braun in his castle laboratory. Leo Graetz was there too. The three men sat down to lunch, but for some curious reason the conversation did not get around to Hertz's great achievement; he was too modest to start on the subject, and neither Braun nor Graetz said anything about it, either. Six months later Hertz remarked about this curious occasion in a letter to his parents, on 14 October 1888: "When I was in Tübingen at the beginning of April, I had ready all the material that has now been published and would have loved to talk about it, but no opportunity arose, since no one else brought up the matter, although by then enough had been published to make a physicist eager to ask questions."[246] Perhaps there was some diffidence on Braun's part, too, a reticence about intruding on a colleague's field.

Yet it is also possible that Braun did not find Hertz's work immediately convincing. A couple of months before Hertz's visit, Braun had written to Leo Graetz in Munich,

I am no longer impressed with Hertz's effect of ultraviolet light on electrical discharges, after having determined recently that the effect as good as disappears in the electrostatic case, when magnesium illumination alters the discharge potential by no more than 2–3%, too low to be measured. H. should have tried that and described it! But it is a dreadfully exaggerated representation. That's just like his elastic floating plate [an earlier paper in which Hertz had rigorously treated the paradox that a floating deformable plate may be prevented from sinking by an *increase* of the weight on it[341]]—the whole thing can be stated in a couple of words, only then everyone would know that a piece of paper with a bulge in it can support a greater weight the more it bulges.[104]

Despite Braun's misgivings, Hertz's observations on the effects of ultraviolet light on electrical discharges were correct. In fact, they were the first reported observations of what became known as the photovoltaic effect, which a few years later helped to revolutionize physics with the introduction of quantum theory.

Braun did not make up for his delay in acknowledging Hertz's achievement until 1889, by which time it had been generally recognized throughout the world. In September 1889, at the Heidelberg meeting of the German Society of Natural Scientists, he congratulated Hertz.

At the time of Hertz's visit to Tübingen, Braun's family already included two children. The first son was born on 1 March 1886 and was christened Siegfried. As promised, Carl Engler was godfather. A daughter, Hildegard, was born on 10 March 1888. The birth of his first child coincided with the arrival of the reprints of his inaugural lecture. In sending them to his friends and acquaintances, he added to each dedication, "And now a happy father, too."[303]

Braun's other experiments during these years were necessarily on a small scale. In accordance with Goethe's dictum, "Under constraint the master first shows plain," he returned to problems that had fascinated him in his youth: rock salt and the electroscope. During the winter of 1886–1887, Braun had investigated the electrical properties (the high resistance) of ordinary rock salt. The results were reported in a paper sent to the *Annalen* on 20 May 1887.[26]

This paper is of particular interest because it contains the first description of the *Braun electrometer* (that is, the electrostatic voltmeter; see figure 10), a particularly rugged and practical version of the electroscope. Such instruments had long been in use, usually in the form of two extremely thin gold leaves, which were hard to handle, expensive, and not particularly sensitive; nor were they calibrated in volts. Braun, with the help of his mechanic Albrecht, replaced the gold leaves by considerably thicker aluminum leaves, improved the configuration, and calibrated the resulting instrument in volts—an achievement of considerable practical importance.[124] It was simple and convenient and presented a low capacitance, properties that insured its continuing employment in physics and engineering for years to come. Albrecht persuaded Braun to exploit the new arrangement commercially and con-

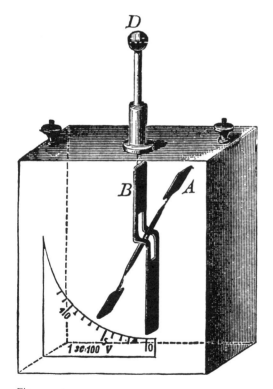

Figure 10
Drawing of Braun's electrostatic voltmeter, 1891.

structed three versions of the instrument, with ranges of 1,500, 4,000, and 10,000 volts. In 1891 Braun published descriptions of the apparats in the *Annalen* and in a science educators' journal.[41]

After nearly two and a half years of work, the new physics institute was finished in October 1888. Drainage problems and the need to control underground streams had delayed the start of construction. Now the local paper could announce, "A tasteful fence with rich decorations surrounds the entire building. Broad sidewalks give the street a metropolitan appearance."[222] Except for this street, Tübingen was certainly no metropolis, though its 13,000 inhabitants could count on two visiting theater companies each week and discussion groups and social events almost every evening. In addition to the Tuesday Club, there was a Wednesday bowling club. This physical exercise did not prevent Braun from putting on some excess weight, a natural consequence of his healthy pleasure in good food and drink. Contemporaries recalled jolly evenings at his house, sometimes enlivened by entertaining caricatures that Braun drew of his guests.[325]

The year 1888 was the Year of the Three Emperors. Wilhelm I died and his son, Friedrich III, already fatally ill, abdicated after ninety-nine days in favor of his son, Wilhelm II. Braun followed these events with great interest, but his main concern was the last major piece of research he was to undertake in the old castle laboratory, a large-scale project that overflowed its bounds until it had spread throughout most of the building.

Metal coils of all sizes hung from the ceiling. Spirals, stretched wires, and electrical connectors all over the place made a shambles of the laboratories. Meters, wire-pulling apparatus, and hissing gas flames completed the picture. In the midst of it all was the master, happy as a child at play, experimenting with a discovery that might make him rich and famous.

The discovery was what Braun called "deformation currents." The idea went back to his Karlsruhe observation that the thermoelectrical behavior of solids was affected by all sorts of gross mechanical changes. Now he looked for electrical currents evoked by changes in pressure at points of electrical contact. He noticed that bending moderately thick metal wires back and forth rapidly produced a measurable thermoelectric effect.[32]

Did this observation relate to thermoelectric currents caused by the frictional heat generated by the bending, or was it perhaps a magnetic effect (the so-called torsion current) caused by the twisting of the wire about its long axis? Braun thought it was neither, but rather a "direct and complete conversion of mechanical energy into electrical energy," a novel phenomenon that would have been of considerable practical and scientific significance. He immediately submitted a two-part paper on his discovery to the grand old man of natural science, Hermann von Helmholtz: "On Electric Currents Produced by Elastic Deformation" and "On Deformation Currents; Particularly As to Whether They May Be Related to Magnetic Properties." These contributions excited considerable interest. One could speculate about novel methods of producing electricity, for instance, from metal coils alternately stretched and compressed by water or steam power—to say nothing of meters or motors using the reverse principle to produce motion from electricity.

Nothing came of it. While the scientific papers were still being reprinted in the technical journals, Ludwig Zehnder, Röntgen's assistant at Würzburg, proved that deformation currents were nothing but the old magnetic torsion currents caused by twisting a wire, as first described by the Irish scientist William Sullivan in 1845.[251] The refutation upset Braun greatly. He had paid particular attention to this possibility, yet his experiments in that direction had consistently yielded negative results. Finally he had to admit that Zehnder was right. In a third paper to the Berlin Academy[35] and in a later paper in the *Annalen*,[36] Braun conceded that there were two ways of producing torsion: either by twisting a straight wire or by coiling the wire and then stretching or compressing the coil in such a way that the two ends were not twisted relative to one another.

That ended all dreams of financial returns. Torsion currents were too small to justify commercial exploitation. Yet this debacle caused no permanent damage to Braun's reputation. For one thing, the iron supports of the adjacent university library and vibrations in the structure of the building had caused magnetic and mechanical artifacts that Braun had erroneously assumed to have been the results of his experiments. This episode was thus the final contribution of the castle-tower laboratory just before Braun moved into his new institute. For a brief period, the experiment had held the stage. For example, Zehnder remarked that his chief, Röntgen, had spent hours "thinking about and correcting the little paper refuting Braun."[252]

The new physics institute was dedicated on 10 January 1889. The ingenious design of the building showed how well Braun and architect Berner had absorbed what they had learned on their ·inspection trip.[306] The building rested on deep foundations of solid concrete and heavy beams to make it as free as possible of vibrations. Instead of the usual tower for experiments involving free fall or pendulums, the designers provided a series of trap doors on three floors. From Strasbourg was taken the idea of inserting small windows in interior walls in such a way that optical experiments could be done over the whole length of the building. The problem of maintaining constant temperatures was solved by the construction of two special rooms with double windows and doors. The room intended for heat experiments had a floor sloping to one point; mercury from broken thermometers would collect at this spot. Doors had no sills, so that equipment on movable tables and racks could be conveniently rolled from one room to another.

Separate rooms were devoted to optics, photography, chemistry, and time measurements, and the director had a private laboratory. The power equipment was of the best. A 6-horsepower engine made by Deutz was installed in a special room. It had the usual belt transmission and it drove an electrical generator made by the Esslingen Machine Works. This special room was the first electrical power plant in Tübingen. (The town's first electrical generator outside the university was not installed until five years later, when the local slaughterhouse was electrified.) The university's generator supplied electricity to a network that extended to almost all the institute's rooms. It also served to charge a large battery, so that the four arc lights (of 600 candlepower each) in the main lecture room could be lit even when the machine was not running.[49] These lights became a local attraction. During evening lectures, when the lamps were lit, institute porter Georg Schurr had to chase away youngsters who wanted to partake of technological progress by climbing all over the ornamental railings outside the lecture room.[222]

"I confess I am only now begining to enjoy Tübingen," wrote Braun to his friend Leo Graetz after the move. "At last I feel as if I am in a proper institute." Evidently students had the same feeling, for the next semester showed a significant increase in enrollment. Although Tübingen was one of the largest German universities of its day, with a mathematics and science faculty that dated back to 1863, physics students always had avoided it because of the miserable state of its laboratories. Of the 1,400 students who were enrolled in 1886, only 86 specialized in mathematics and natural science. After the new building was completed, this fraction grew very quickly.[218]

Working under Braun were two *Privatdozenten* (unsalaried lecturers) and an assistant. *Privatdozent* Karl Waitz gave the lectures in theoretical physics, normally the assignment for a second full professor. Dr. Waitz had also come to Tübingen from Karlsruhe, where he had been assistant before Braun had been there. He was knowledgeable in meteorology and astronomy, and he developed a good working relation with Braun, who was only three years his senior. Waitz's lectures were "extraordinarily well prepared and instructive,"[135] as the government finally recognized by appointing him associate professor in 1891. He showed his gratitude by remaining at Tübingen until his death in 1911.

The second *Privatdozent* and chairman of the institute was Dr. Otto Schumann; he left in 1890 to go into industry. He settled finally at Krupp's, where he remained until his early death in 1898.[156] Walter Negbaur became assistant in the spring of 1890. In the winter 1891–1892 semester Mathias Cantor, a dark-haired Austrian, was named assistant. The young man, "most susceptible to feminine charms and somewhat uneven in his scientific work,"[132] remained with Braun for thirteen years.

Another long-time collaborator who first made Braun's acquaintance in Tübingen was Jonathan Zenneck, a minister's son from Ruppertshofen in Swabia. Zenneck matriculated at 18, intending to become a mathematics and science teacher at a secondary school. For nine semesters he attended Braun's lectures on experimental physics and participated in the laboratory exercises and physics colloquium that Braun had reintroduced.

Early in 1890 the government of Württemberg offered an interesting opportunity for experimentation. A trial drill hole had been sunk to test for coal near Sulz in the Neckar valley. The drill had reached a depth of 900 meters without success. The Director of Mining, Dr. von Baur, did not want the effort to go to waste altogether.[46] Would Professor Braun be interested in using the deep shaft to measure the increase in temperature in the earth's interior? Such measurements were not only of theoretical and scientific interest but also of technical importance, since they showed how far below the surface of the earth miners might be able to work. Accordingly, the government appropriated the necessary funds and Braun accepted the undertaking. An outpost of the physics institute was set up in the village of Sulz, an hour and a half by train from Tübingen, and Braun and Dr. Waitz took turns in running it. A special thermometer on a steel wire was lowered into the muddy water of the bore hole. It took half a day to lower the thermometer and another half a day to bring it up. The distance was six times the height of the cathedral tower in Cologne. Temperatures at various depths were registered by a measurement of the amount of mercury driven out of an open tube.

The measurements took three weeks; the evaluation of the results, one year. In 1892 Braun and Waitz reported that the temperature appeared to rise by 1°C with every 24.10 meters of depth, a lower value than had been reported elsewhere.[46] These measurements were the most peculiar scientific task that Braun had undertaken and one of only two to result in a coauthored paper.

With new assistants and a new building, it was possible to think of interests in physics other than the direction of the institute. Walter Negbaur developed a new method of measuring torsion with an accuracy of 0.1 percent. He came to this result through the director's work on deformation currents and on the elastic aftereffect. In another paper, Negbaur extended Braun's earlier work by developing a method of determining potential differences across the interface of two solutions.[237]

Braun helped his assistant in his work on rotation, improving an instrument for the measurement of length (the so-called comparator, figure 11), so that Negbaur could use it also to measure

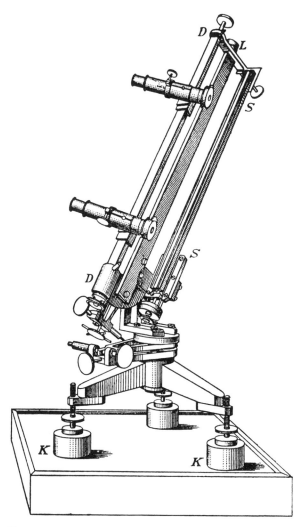

Figure 11
Braun's comparator for measurement of very small distances.[37]

horizontal connecting lines. He described this instrument in the fall of 1890 in the *Annalen*.[37]

In the same volume Braun published a paper "On Drop Electrodes."[38] The passage of a drop of mercury through diluted sulfuric acid gives rise to an electrical current, a phenomenon that Braun had investigated while in the old institute "for his own information." In April 1890 the Parisian physicist Henri Pellat then derived a law that no electrical potential exists between a metal and its salt solution.[342] Braun took issue with Pellat, for he himself had observed such a potential. The controversy dragged on in the *Annalen* until January 1892, without a definite result.[47]

Braun's extraordinary memory allowed him to relate a phenomenon that he had noted during his experiments with drop electrodes to an observation made by Theodor von Grotthuss "quite incidentally, as described in Gilbert's *Annalen* for 1819." If the current through certain electrolytic solutions is made to pass through a narrow slit in a dividing wall, the slit begins to act as a third electrode: metal particles are deposited there and bubbles of gas form.[343] Braun was so fascinated by this phenomenon that he made it the subject of his first major investigation in the new physics institute. He experimented extensively to determine in which materials and under what circumstances the phenomenon took place. He reported his observations in a paper at the Berlin Academy of Sciences on 20 November 1890, "Observations on Electrolysis," and on 15 November 1891 he described details of the process in the *Annalen*.[44] He designated the phenomenon *electrostenolysis,* after the Greek work *stenon* (narrowness). Braun found that the phenomenon sets in as soon as a certain current intensity is reached, even if a piece of old tracing paper is used as a dividing wall. He tried to account for the phenomenon by the composition of the solutions, coupled with certain molecular processes. He also saw an analogy with spark discharges, which also take place only above a certain threshold potential. In a third paper, "On Electrocapillary Reactions," Braun compared electrostenolysis with similar processes being investigated in Paris by Edmond Becquerel,[45] but again he was unable to draw any conclusion. Even today this curious phenomenon has not been explained.

In 1891 Braun demonstrated the advantages of long-distance electrical transmission at the Tuesday Club. In the summer of 1891 the engineering firm of Oskar von Miller for the first time had transmitted electrical power over long distances. They had coupled an electrical generator to the turbine at the main installation near Lauffen on the river Neckar, transformed the generated voltage to 10,000 volts, and had then sent it a distance of 175 kilometers to Frankfurt. "Power from Lauffen" fed the 1,000 lamps at the International Electrical Exhibition, as well as a pump that operated a 6-meter artificial waterfall.

Some laymen thought that a single waterfall, like the one on the Rhine at the Swiss-German border, would provide enough power to light all Europe. Others feared that the high-voltage overhead lines would lay the land to waste by a shower of lightning bolts.

Just the task for Ferdinand Braun! In a single concise lecture intended for the general public he developed the subject of electricity from basic principles to speculations about the future. Each item was described before being demonstrated and appeared at the right moment as if by magic from behind a screen. His lecture was such a success that he was asked to repeat it the next day. Despite the relatively high price of admission (1 mark for seats and 0.50 mark for standing room), the lecture hall was jammed. The *Tübingen Chronicle* noted that this turnout reflected "the interest our age takes in the development of electricity."[222]

Braun was encouraged to publish this lecture as a pamphlet, *On the Transmission of Electrical Energy, Particularly Alternating Current*. The booklet, thirty-six pages long, was brought out by the H. Laupp Bookstore in Tübingen.[48] The notes accompanying the text show that Braun had no doubt at that time that Tesla's motor was "the fundamental development of the whole field . . . incontrovertibly so, even though some of these things had been known previously." That was by no means the general view at the time. The engineer Dolivo-Dobrowolsky was widely thought to be the discoverer of the underlying principle. Besides the "essential importance" of Tesla's invention, Braun particularly admired "the very detailed description in the patent claim."[48] He could not suspect what difficulties Tesla's habit of making detailed patent claims would give him seven years later.

The beginning of the age of electricity ought to have been an excellent opportunity for the partners in the Braun Electrometer Works of Tübingen to grow wealthy. The high voltages used to distribute electrical power created a great demand for instruments by which these voltages could be measured. The Braun electrometer was just such an instrument. Customers were given reprints of Braun's papers on the electrometer, in which the instrument's merits as a high-voltage device were described. But the small company found it hard to compete for a share of the market, especially since it made an attempt to follow Braun's maverick suggestion that a purely utilitarian instrument should be manufactured. At the International Electrical Exposition in Frankfurt Braun, together with Quincke and Kohlrausch, had criticized German industry for putting too much effort into the highly polished hardwood boxes used to house the instruments and had complained that "few people were intelligent enough to avoid this pitfall,"[242] that is, prizing the instrument for its container.

The Tübingen company tried to imitate foreign suppliers by manufacturing an inexpensive housing, but the idea did not catch on in Germany. Braun's opponents in the Frankfurt discussion (with his brother's partner Eugen Hartmann as their spokesman) maintained that anyone who wanted to sell industrial products must continue to pay great attention to the external "finish." The Tübingen concern failed. Albrecht continued to putter around with the production of the Braun comparator, while Braun argued in the *Annalen* with the chemist Ostwald, who had taken issue with one of the observations in Braun's paper on drop electrodes. Elsewhere, he took up several columns in the *Electrotechnische Zeitschrift* with complaints that, in spite of his work in 1878 and 1882 on the conversion of chemical into electrical energy, most textbooks of electrical technology and even of physics were still perpetuating false ideas.[42]

Braun's ingenuity as an experimenter came to the fore once again in the supervision of his first doctoral student in Tübingen, his assistant Mathias Cantor. Cantor wanted to determine the constant of capillary attraction by measuring the adhesive forces demonstrated when a ring floating on the surface of a liquid is lifted off. He started by connecting the ring to a scale in order to measure its apparent weight. But with this setup he could not satisfy the assumptions made in his theoretical analysis. The experiment succeeded after Braun suggested that first the ring should be secured and then the vessel containing the liquid attached to the scale.[253] This configuration (which was a success) derived from one of the experiments in Braun's book on science for young people,[6] in which the father presents the group of children with this well-known teaser: Does a glass of water grow heavier when you stick a finger into it? He shows them how to construct a simple scale by placing various weights on a glass attached to a tree branch and using a marked stick to register the branch's deflections. The method was one example of Braun's *physica pauperum*: the string-and-sealing-wax kind of physics. The art of devising demonstration experiments using the simplest means at one's disposal had aroused Braun's interest when he was quite young. In Strasbourg he had lectured on this subject to his department colleagues.[325] His old teacher Quincke held the same point of view. After visiting him in Heidelberg, Braun reported with enthusiasm how Quincke had contrived a substitute for the Thomson quadrant electrometer. "The most expensive parts of it were a glass bottle and a mirror; the remaining parts were a little box and a well-lubricated bearing surface made of strips of mirror glass affixed with sealing wax; finally, a little box cut into four quadrants. The electrometer was as good as the original instrument and more convenient to use."[78]

Despite some misgivings,[104] Braun had undertaken to prepare the chapter on thermoelectricity for the *Handbuch der Physik* edited by his colleague Winkelmann. That gave him an opportunity for a practical review of a group of topics that he had previously approached theoretically, in Karlsruhe. Noting that various metal pairs used as thermoelements produced different potentials when heated by a given amount, Braun hypothesized,

Let us imagine the thermoelements as consisting of generators of molecular dimensions, with many such generators in each element and all working with the same efficiency. Differences in electromotive force arise because some elements contain relatively many generators and others only a few. We are thus led to the assumption that the molecules of a metal, when considered as thermal generators, are not all equally effective.

Braun published this hypothesis in June 1893 in a paper entitled "On the Physical Interpretation of Thermoelectricity," in the *Annalen*.[50]

In October 1893 he completed a treatise, "On Continuous Conduction of Electricity through Gases," a topic that had occupied him now and again since 1874 but for which he had found no occasion to publish anything.[51] An occasion now arose: a publication based on the dissertation of a Belgian doctoral candidate, Alexandre de Hemptinne in Leipzig, who later became a professor at the University of Louvain. He had obtained experimental results that Braun was able to confirm from his own studies in Würzburg on current conduction through gases on the point of explosion; in Karlsruhe on the conductivity of gases at high temperatures; and in Tübingen on the conductivity of gases being formed by chemical combination while being blown through a glowing tube.

At the beginning of the 1893–1894 winter semester, Tübingen got its first woman student, the energetic Countess Maria von der Linden, a hearty Swabian who came to study zoology and later achieved professorial rank when she became director of the Institute of Parasitology at Bonn. The Tübingen faculty members were not sure how to handle the situation. Some solemnly warned the students beforehand that a woman student would join them on the morrow. One professor offered her the use of his office, so that she would not have to remove her coat in front of all the other students. Braun doubtless took it all in his stride. We know that eight years later, when a botanist colleague turned down two women students who wanted to study biology, Braun was reported as saying, "I don't know what he has against women students. I wouldn't mind seeing a pretty girl in the first row of *my* class!"[328]

Braun had no inkling that his lecture for 13 November 1894 before the Tuesday Club, on "Heat and Light," would be his last before that body. It had been a dreadful year for German physics. Heinrich Hertz, 36, had died in Bonn on New Year's Day (see figure 12). August Kundt, 57, died in Berlin in April. And Hertz's teacher Hermann von Helmholtz, 73, died in September in Berlin. Two top positions in Berlin and an important chair in the provinces

Figure 12
Heinrich Hertz (1857–1894).

stood vacant. Braun may have guessed that he would be involved, but he surely did not reckon that the "landslide of 1894" would carry him to a high position in the most important center for physical sciences, after Berlin, in all Germany: Strasbourg.

Strasbourg had developed into an important center not only because of its modern institute, but also because of the consistently high quality of its faculty; it had become known as a source of talent. When Kirchhoff died in 1877, Kundt was called from Strasbourg to Berlin to replace him. Again, when Helmholtz died in 1894, the choice fell on Kohlrausch in Strasbourg to replace him as head of the National Physical Laboratory. What it meant to Ferdinand Braun when, in the winter of 1894–1895, he was asked whether he would be interested in replacing Kohlrausch, needs no comment. He accepted the honor, though not without a certain regret at leaving Tübingen, where he had spent many years of personal and professional happiness.

Tübingen was also dismayed by the news that Braun had accepted the Strasbourg chair. He had been a very popular lecturer and club member. Testimonial followed testimonial, some of them in a humorous vein. The Tübingen Mathematics Club presented him with a table of logarithms carried out to only *one* significant figure, an allusion to his skill in computation.[135] The bowling society of the Wednesday Club dedicated a little verse to their departing member:

Und Braun, der edle Physiker, ist reich an Anekdoten.	And Braun, the noble physicist, The witty storyteller,
Hat er 'nen . . . guten Witz gemacht,	Likes to tell tales—and if they pall,
'ha, ha' er wie 'ne Bassgeig' lacht	His bass profound is heard withal;
Dann füllet er mit Wohlbedacht den Kegelplatz mit Toten.	Then calls he for his bowling ball And knocks 'em in the cellar.

The Tübingen contingent arrived in Strasbourg in March 1895. It comprised Professor Ferdinand Braun, 44; his wife Amalie, 36; his children Siegfried, 9, Hildegard, 7, Konrad, 3, and Erika, nearly 2; a maid; and a dog. Braun's assistant, Mathias Cantor, 34, also came along. They were welcomed by Associate Professor Emil Cohn; *Privatdozent* and First Assistant Adolf Heydweiller; Second Assistant Dr. Josef Wellstein; Laboratory Assistant August Klughertz; the porter Karl Sittler; and 179 students of the faculty of science.[203]

Emil Cohn had been assistant at Strasbourg during Braun's first stay there. He was four years younger than Braun and very reserved; it was difficult to get to know him.[135] His book, *The Electromagnetic Field*, on which he was working at the time, and his way of expressing Maxwell's equations so that they could be transformed into any system of units received recognition in professional circles. The porter, Karl Sittler, was likewise an old acquaintance, "most meticulous and diligent but not excessively bright."[135]

The first assistant, Dr. Heydweiller, stayed only one more semester before accepting an associate professorship in Breslau; later he became professor in Münster and in Rostock.[156] He was replaced by Mathias Cantor. Another assistant's position was found for Jonathan Zenneck, Braun's student in Tübingen. Zenneck originally had studied zoology, obtained a doctorate, worked for a time in London in that field, and returned to Germany for his tour of military duty. When he received the offer to join Braun's staff in Strasbourg, Zenneck was only 25, and he "gladly accepted, in order to learn physics, which I had always liked, even as a student, but expected eventually to return to zoology." Actually Zenneck remained a physicist and ultimately achieved a high reputation in the field. At the last moment there was a temptation to work in zoology: he was offered an assistantship by Eimer, the zoologist who had supervised his dissertation at Tübingen, fourteen days after Braun's offer. Zenneck wrote to Braun asking to be released, but Braun had already pocketed Zenneck's acceptance. "I already have an assistant," he wrote Zenneck; "let Eimer look for another one."[135] This somewhat unbending attitude must be viewed in light of the circumstances. As the number of students in the natural sciences was small and most of them were drawn to the more lucrative positions as science teachers in secondary schools, it was difficult to find assistants.[135]

Braun engaged Hermann Meyer as a general mechanic and custodian. Meyer was a former railroad employee who had applied for the job because he was so attracted by the work of the institute. Braun found him most satisfactory and made arrangements for Meyer and his family to live in. Meyer never ceased to astonish the staff members, whose ideas he often realized in a way that was far superior to their original concept. His precision and diligence were much appreciated by Braun.[124]

Tesla's generating station to light the Strasbourg railroad station had been one of the first in Germany, and now in 1895 the city was among the first to replace its power station by an ac source. The progressive physics institute asked to be connected to the new ac supply from the start. The physics department was fascinated with the new-fangled devices that came with the ac installation and resolved to learn as much as possible about them, which in those days was not much; even most electrical engineers understood ac only imperfectly. It all fell into place when Prof. Cohn spoke the magic words: "rotating magnetic field" and "induction currents in the armature." A little later Cohn gave a lecture at the institute about the new system. Sitting in his audience were Strasbourg's electrical engineers, eager to understand the system they had just installed.

It is a property of ac that it passes through two maxima of opposing polarities per cycle at a rapid rate, 25 times per second in some of the early systems or 50 maxima per second. The eye does not normally perceive this rapid variation, except under certain conditions, such as when an intensively illuminated object is rapidly moving back and forth. (This objectionable flicker later led to the adoption of higher frequencies, 50 cycles in Europe and 60 cycles in America.) Braun almost caused a public incident one evening near the railroad station when someone asked him to explain the new system. He had walked over to one of the arc lights that had just been converted to ac and rapidly moved the extended fingers of his hand back and forth to demonstrate the flicker phenomenon. A sizable crowd quickly collected. Everyone had to try the experiment. Some were more successful than others, who then tried to move their hands even quicker. The crowd grew as more and more citizens joined in the strange rite. A contemporary account reports that policemen nearly intervened in what they assumed to be an outbreak of mass madness.[320]

On New Year's Day of 1896 Ferdinand Braun, alone in the deserted institute, read one of the strangest accounts he had ever encountered—the description of an almost unbelievable phenomenon.[135]

If we pass the discharge from a large Rühmkorff coil through a Hittorf or a sufficiently exhausted Lenard, Crookes, or similar apparatus, and cover the tube with a somewhat closely fitting mantle of thin black cardboard, we observe in a completely darkened room that a paper screen washed with barium-platino-cyanide lights up brilliantly. . . . It is soon discovered that all bodies are transparent to this influence. . . . Paper is very transparent; the fluorescent screen held behind a bound volume of 1000 pages still lighted up brightly. . . . Fluorescence was also noted behind two packs of cards. . . . Thick blocks of wood are also transparent. . . . If the hand is held between the discharge tube and the screen, the dark shadow of the bones is visible within the slightly dark shadow of the hand.[344]

Braun was reading a reprint that Röntgen in Würzburg had just sent to several colleagues of his acquaintance. It was his preliminary communication, "On a New Form of Radiation," rays that were to become known in Germany as Röntgen rays but that Röntgen himself called x rays. No one had ever heard of rays capable of passing through matter. Even Braun remained skeptical at first.[116] When Zenneck came back after the Christmas vacation, Braun handed him the reprint and remarked, "Röntgen has otherwise always been a sensible man—and it isn't even carnival time yet."

But the Strasbourg institute was soon seized by the general fever of excitement over what was surely one of the most exciting physics discoveries of all time. Zenneck reports,

The rays were fantastic. Everywhere there were lectures about them. Braun, too, gave one—I might say with my cooperation, since I had to pump out the cathode-ray tube in which the Röntgen rays were produced—which, given the air pumps we had then, provided some healthy exercise. This tube yielded only quite weak Röntgen images; after looking at one for half an hour, an impressionable soul might discover a suggestion of the bones of a hand.[135]

Röntgen's discovery opened the gates for a flood of research all over the world. It is a quirk of fate that Ferdinand Braun, as skeptical in this case as when he first heard of Hertz's discovery, was to produce within the year one of the mightiest waves of this flood.

Braun's mother Franziska Braun died on 5 February 1896 in Fulda; she was 80. The family's ties to Fulda were severed with her death. Two years earlier, her second-oldest son Ludwig had died in Bremen at 53, after a brief and severe illness.

The task of keeping track of Röntgen's discovery for the Strasbourg Physics Institute fell to a postgraduate student from Russia, Alexander Eichenwald.[135] Assignment of such an important task to a foreigner was nothing unusual in Strasbourg, even at a time when nationalism was at its height. From the first, the new German University of Strasbourg drew more foreign students than any other in Germany, precisely because it was free of the traditions, often rooted in nationalistic fervor, that prevailed elsewhere.[204] They were attracted above all by this new type of "working university." Nearly one tenth of the 1096 students in the 1896–1897 winter semester were foreigners, including 24 nonwhites. Most of these foreign students specialized in natural sciences. Word had got around that Strasbourg had the best equipment and the best teachers in that field. The largest group of foreign students were

Russians, followed by the Swiss and Americans. Sons of well-to-do families who wanted to take advantage of the combination of excellent instruction and cosmopolitan atmosphere sat next to students who were driven by the sort of thirst for knowledge that characterizes the youth of developing nations.

Alexander Eichenwald came from St. Petersburg and had served as graduate engineer for the Russian railroads and as city sanitation engineer in Kiev before he decided on an academic career.[156] He had come to Strasbourg when he was 31, in 1895, and had taken over from Dr. Wellstein, who had been appointed *Privatdozent* in mathematics and later became professor in Giessen and in Strasbourg. Eichenwald's doctoral dissertation at Strasbourg, which he completed in July 1897, was an extension to conductors of Cohn's application of Maxwell's theory to nonconductors. In 1897 Eichenwald returned to Russia, where he subsequently became director of the Physics Laboratory at the Technical University of Moscow, professor at the University of Moscow, and chairman of the Russian standards commission. When the Soviet government sent him to Germany in 1920 he chose not to return and lived out his life in meager circumstances, first in Prague and later in Milan,[135] where he died in 1944 at the age of 94. In 1932 he wrote an article about Braun for the Russian technical journal *Elektrichestvo*.[117]

Eichenwald's duties as "radiation specialist" made his stay in Strasbourg an exciting one. "The new rays of Prof. Röntgen" overshadowed all other scientific and engineering discoveries of the time. The general public was particularly fascinated with the medical possibilities of x rays.

In consequence, other discoveries of comparable significance made during 1896 suffered. The French physicist A. H. Becquerel, for example, showed that radiation emanated from uranium; that is, he had discovered radioactivity. An important publication by the Dutch physicist H. A. Lorentz made a substantial advance beyond Maxwell's theory. A young Italian, Guglielmo Marconi, was experimenting with wireless telegraphy. And in France, the brothers Lumière were applying elements of physics, physiology, and chemistry to their epoch-making *cinematograph*.

The four small papers that Braun completed during 1896 were also overshadowed by Röntgen's work. He was not successful in his "Experiments to Demonstrate Electrical Surface Conductivity in a Preferred Direction," in which he tried to show that the orientation of new particles in a growing crystal is caused by an "electrical orientation" that is present in the liquid from which the crystal is grown.[53] He had better luck with observations that led him to conclude that the electrical properties of a thin liquid layer over a solid were continuous with those of the solid itself. The investigation involved passing a current through plaster disks in a humid atmosphere over a period of four days.[54] Braun wrote that he found these results particularly interesting because they afforded "a welcome and long-sought confirmation of concepts about the nature of the liquid state" that he had formulated for himself and to which he "presently hoped to return." (That "presently" took 20 years.)

A third paper dealt with the old problem of what happens to the electricity associated with a water droplet that is vaporized. In his "Investigation of the Conductivity of Electrified Air," Braun sent a stream of air through the brush discharge of an electrical generator and showed, by means of wire gauze, that the air became electrically charged.[55] But this result, obtained by conventional means, created no great excitement: before it had reached print, everyone had become aware of the fact that x rays could discharge a charged body. The same method provided the answer to the question whether electricity emitted by a glowing point must be carried by solid particles. Before Braun had a chance to make this experiment, it became clear that "the answer to this question was no longer of interest inasmuch as Röntgen had shown that air can become a conductor upon irradiation with his rays."

Braun's fourth paper of 1896, "A Magnetic-Current Experiment," reported an original experiment.[56] Rather unwieldy arrangements had been used until then to show how the motion of a magnetic field in a stationary conductor induced electricity in a neighboring wire. Braun wound a coil of iron wire about a straight copper wire and placed the iron-wire coil between the poles of an electromagnet. Changing current in the magnet thus led to changes in magnetic polarity in the iron wire, which in turn induced a current in the copper wire. Braun submitted all four papers of this "Röntgen year" to the Göttingen Academy of Sciences on 20 September 1896.

Radiation experiments were soon taking place in the Strasbourg Physics Institute, too. Large wooden boxes marked "Caution! Glass" started arriving from Bonn, home of the best-known glassworks, "Franz Müller—formerly H. Geissler." One such box, which arrived during the winter of 1896–1897, contained a tube about 50 centimeters long with a cone-like flare at one end and a faintly shimmering deposit on the inside, as well as various metal structures. The entire staff came down to watch as the instrument was carefully unpacked. Franz Müller had followed the design to the millimeter—an accomplishment that everyone in the room appreciated, for at that time every physicist was his own glass-blower and mechanic.[135] Only so-called final arrangements, such as the tube described above, were acquired from commercial suppliers.

The tube was quickly set up on a workbench, electrical connections were made, and the vacuum pump was started. When the electrostatic high-voltage generator was turned on, a faint glow became visible at the screen end of the tube. More pumping. On a second try, a distinct spot of light became visible on the screen. Braun brought the magnetic-field coil closer to the tube and applied alternating current. The spot of light became a wobbly line. Braun flicked the rotating-mirror viewer into motion and observed the line through it. Finally, he rose, yielded his place to his assistants, and invited them to meet the alternating current of the Strasbourg generating station "in person."

If they were conscious of participating in a historic moment, they were right. Ferdinand Braun had also caught the "radiation fever" that followed on Röntgen's discovery, but his critical mind recognized that not much remained to be done in the direction of finding new properties of x rays or in looking for entirely new kinds of rays at random, which was what most of his contemporaries were doing. His intuition proved to be correct. Röntgen had investigated his discovery so exhaustively that little new was to be added for years.

Braun had no taste for indulging in random speculations. Instead, he characteristically took up his usual quest for "the simplest means." He went back to the elegant experiment that the mathematician and physicist Julius Plücker had made in Bonn in 1858— the basic configuration for all radiation experiments, including

Röntgen's. It consisted of a glass tube with closed ends through which two wires penetrated into the inside from opposite ends. All sorts of colorful phenomena could be observed when the device was filled with various gases at varying pressures and a current passed through it. When the tube was evacuated, a glow appeared near the negative electrode, the cathode. Plücker concluded that the glow was caused by a new kind of rays, which he called cathode rays.[345]

The tubes had been constructed in Bonn by Heinrich Geissler, who later went into business making them. His "glow tubes" were exhibited as curiosities at country fairs, long before their descendants, neon and fluorescent lights, were to carry all before them in illumination technology. Plücker established that cathode rays travel in straight lines and that they cause a glow whenever they hit the tube wall. He discovered further that they can be deflected by a magnet. During the following decades these properties were investigated by a number of scientists, including Johann Wilhelm Hittorf in Münster, Sir William Crookes in London, and Heinrich Hertz and his assistant Philipp Lenard in Bonn. None of the investigators thought of *using* cathode rays for anything, no one, that is, except Ferdinand Braun.

Röntgen's results also depended on evacuated glass tubes and cathode-ray luminescence, which induced Braun to take a closer look at the cathode rays themselves and so led him to the invention of the cathode-ray oscilloscope—the heart of which, the cathode-ray tube, is still known in Germany as *Braunsche Röhre* ("Braun's tube"). Its most familiar embodiment—perhaps the most familiar electronic device of all time—is the television picture tube.

What was new about the tube that Braun first demonstrated on 15 February 1897? By now the cathode rays were familiar, as were their ability to illuminate a phosphorescent screen, the method of limiting their emanation from the cathode into a thin beam by means of an aperture, and their deflection by a magnetic field. What had been lacking was a constructive mind that could assemble all these elements into an effective whole.

In Braun's tube (figures 13 and 14) a potential is applied between the cathode K and the anode A. The cathode rays leave the cathode and travel initially in a direction perpendicular to the cathode K, pass through a hole in a thin disk C, and are gathered there into

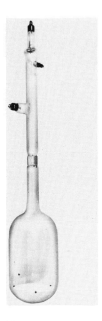

Figure 13
Early model of Braun's tube, now in R. McV. Weston's collection in the United States.

a narrow beam, which strikes the collecting screen D, where a spot of light is produced. An electromagnet (not shown in figure 14) surrounds the thin neck of the tube. The magnetic field from the electromagnet deflects the beam up or down, depending on the direction of the current. Thus in this first version of a cathode-ray oscilloscope the light spot moves only vertically. Until Zenneck added a method of deflecting the beam horizontally in 1899, oscillations could be observed only by means of a rotating mirror, which effectively introduced a second dimension. The mirror was placed at E outside the tube. As it rotated, the viewer saw the vertical line expand into a two-dimensional curve (figure 15; see also figures IX–XI[297] in appendix B).

Braun had begun work on the tube in the summer of 1896 and continued for several months.[57] Franz Müller, the successor of Heinrich Geissler in Bonn, manufactured several versions to Braun's specifications. Not until the winter of 1896–1897 did a satisfactory design emerge. It was described in the issue of the *Annalen der Physik und Chemie* dated 15 February 1897 under the

12. Ueber ein Verfahren zur Demonstration und zum Studium des zeitlichen Verlaufes variabler Ströme; von Ferdinand Braun.

1. Die im Folgenden beschriebene Methode benutzt die Ablenkbarkeit der Kathodenstrahlen durch magnetische Kräfte. Diese Strahlen wurden in Röhren erzeugt, von deren einer ich die Maasse angebe, da mir diese die im allgemeinen günstigsten zu sein scheinen (Fig. 1). K ist die Kathode aus Aluminiumblech, A Anode, C ein Aluminiumdiaphragma; Oeffnung des Loches $= 2$ mm. D ein mit phosphorescirender Farbe überzogener Glimmerschirm. Die Glaswand E muss möglichst gleichmässig und ohne Knoten, der phosphorescirende Schirm

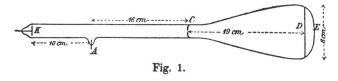

Fig. 1.

Figure 14
Title, descriptive paragraph, and drawing describing cathode-ray oscilloscope.[57]

title "On a Method of Demonstrating and Studying the Time Dependence of Variable Currents."[57] At the same time Braun reported six lesser investigations resulting from the application for which it originally had been intended, as a fast-acting indicator and instrument of observation.

The first of these reports, "Current Waveforms," described the curve of the current delivered by the Strasbourg generating station. Its 50 cycles per second has remained the world standard (except for North America) to this day. The purity of its sinusoidal form, which had so surprised Braun, was confirmed by a tuning-fork comparison that Zenneck had carried out at Braun's request. But the ac produced by an induction generator bore only a slight resemblance to a sine curve. It was in the course of that investigation that the true research value of the cathode-ray oscilloscope first became apparent, for it made it possible to obtain results that could not be obtained in any other way. It was found that the

1896–1897

91

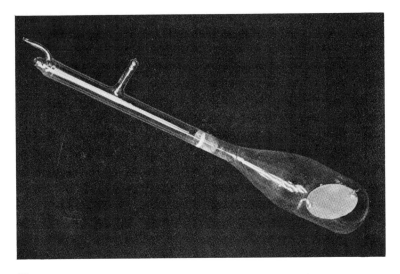

Figure 15
One of the first thirty: Braun tube with skewed display screen, now in
Deutsches Museum in Munich.

iron core of the induction generator became demagnetized with
each change of polarity when the secondary circuit was open, but
retained its magnetization when the circuit was closed. The cathode-ray oscilloscope also facilitated the understanding of related
phenomena, for instance, the closed curves (Lissajous figures) produced by phase differences between two ac signals, and the determination of the velocity of propagation of magnetic excitation
in iron.

Zenneck, who had supplanted Eichenwald as the institute's
"radiation expert," was most enthusiastic about the new device.
He foresaw that it would be possible one day to view current and
voltage waveforms varying at millions of oscillations per second.
"It was just what I had long wanted," he wrote, "an instrument
with which one could *see* what was happening in electrical circuits.
In the days that followed I made it a sort of game to find as many
applications as I could."[133]

One application proved a bit embarrassing. During a lecture
Zenneck demonstrated that the frequency of the Strasbourg generating station did not always remain constant at 50 cycles per
second, but varied between 48 and 52. The engineers in charge

of the station were not too happy about that, but on the whole they tended to regard the new indicator as harmless (although they sat up when they saw that the tube could display their alternating current's harmonic content). They felt that the customers should be grateful to receive current not only at the agreed-upon frequency of 50 cycles per second, but also at an even higher frequency without additional charge.[135]

Publication of the description of the cathode-ray oscilloscope marked a personal anniversary for Braun. It fell fortuitously on the twenty-fifth anniversary of his doctorate.

7

THE FOUNDING OF FIRMS AND GREAT
DISCOVERIES (1897–1908)

From July to October 1897 Braun made a trip to North America. He had been invited to attend the sixty-seventh annual meeting of the British Association for the Advancement of Science,[58] held in Toronto on 18–26 August 1897. This was only the second time that this prestigious organization had met outside the British Isles (the first had been in Montreal in 1884).

In the nineteenth century Great Britain led the world in natural science. As a result, each meeting of the British Association was an important event, and for a foreign scientist to be invited was an uncommon distinction. Braun and his Hanover colleague Carl Runge had been chosen to represent German science at the 1897 meeting. Braun's paper was assigned to Section A (Mathematical and Physical Science), on whose committee served such renowned physicists as Lord Kelvin, Oliver Lodge, and the Americans A. A. Michelson and Simon Newcomb.

The organization, program, and the social events accompanying the meeting were impressive. The city and the University of Toronto, as well as the Dominion government, had been preparing for more than a year to welcome the 1300 participants. During the weekend of August 21–23 Braun joined some 400 other scientists in an excursion to the picturesque Lake Muskoka.[303]

The meeting lasted five days. In Section A alone, more than 60 papers were delivered and discussed. The lunch break sometimes did not come until midafternoon; even so, many important points were left undiscussed. The reporter for the American journal *Science,* Professor Hugh Callendar of Montreal, said of S. P. Thompson's demonstration lecture on cathode rays that "under the hurried conditions of the section meeting, it was not very easy to follow and to observe these several effects."[259] Braun's presentation on August 24 did not fare much better. There were parallel sessions on electricity and on general physics. The electricity session was devoted to electrical waves and oscillations. Three papers dealt with methods for making electrical oscillations visible—a problem made urgent by the increasing utilization of alternating current.

According to Callendar, "Professor Braun showed how, by the action of a magnet on a small pencil of cathode rays directed on to a fluorescent screen, it was possible to demonstrate the form of the wave, either by developing the oscillating line of light into a curve with the aid of a revolving mirror or by using a second magnet to give the corresponding Lissajous figure."[259] The actual summary, which appears in the British Association's Report for 1897 in incongruous German, reads *in toto*:

A cathode ray is deflected in a magnetic field produced by an alternating current. The spot produced on a fluorescent screen by the cathode ray thus executes oscillations that can be analyzed in a revolving mirror. If two mutually perpendicular fields act on the cathode ray, Lissajous figures are produced that make it possible to demonstrate and measure phase shifts between the fields owing to self-inductance, capacitance, polarization, etc. No inertia of the cathode ray has been found; it follows even the oscillations caused by the discharges of Leyden jars. But no magnetic effect of light rays has been found. A magnet placed so that its axis is parallel to the cathode ray broadens it into a figure such as would be produced if the cathode ray were a movable conductor; but whether this figure is the result of an extraordinarily rapid rotation or whether it is at rest remains undetermined. It is of interest that the earth's magnetic field is sufficiently strong to deflect the end of the cathode ray, so that it may prove to be useful for at least approximate determinations of inclination.[58a]

That was the American debut of Braun's tube, a device whose numbers on the American continent alone were destined to run into hundreds of millions, in television receivers, oscilloscopes, radars, and instruments. Yet at the time Braun's invention did not cause a great stir: "The frequency of the coil used was not, however, suitable to show the effects in a sufficiently clear manner to the audience," wrote Prof. Callendar in *Science*.[259] He seemed more impressed by a mechanical device demonstrated by Prof. Edward B. Rosa of Wesleyan University in Middletown, Connecticut. Rosa had turned a galvanometer into an electrical curve tracer by connecting it to a rotating-drum recorder on which a pen drew "very remarkable" details. The arrangement devised by Rosa, who later became chief physicist of the National Bureau of Standards in Washington, had the disadvantage that it could not follow oscillations faster than 30 cycles per second. It was no real competition for Braun's tube, but under the unfavorable circumstances—the large audience, insufficient darkening of the room, frequency mismatch, and insufficient vacuum—even professional observers could scarcely be blamed for failing to recognize the superiority of Braun's cathode-ray tube over electromechanical recording systems.

The trip to North America became particularly memorable because of an offer from the Canadian Pacific Railroad to the conference participants of a round trip all the way to the Pacific Coast for only $70.[258] Braun was quick to accept this opportunity to learn more about the New World and its people. As soon as the meeting was over he rode the length of the Canadian border to Victoria in British Columbia, with a side trip to Yellowstone National Park.

His impressions of this 9,000-kilometer journey were recorded in a large number of drawings and watercolors.[303] On the return trip alone, from Victoria via New York to Genoa, he painted sixty tiny watercolors. Their accompanying comments, which are like diary entries, record his impressions and experiences. He drew a prairie fire in the plains and the untamed Kananaskis waterfalls in the Rocky Mountains. There are drawings of the monotonous salt flats in North Dakota's Missouri Coteau, lonely farm houses in the Canadian wheat country, the prairie chicken region near Winnipeg, and views of lakes large and small—a handsome depiction of the characteristic North American landscape of the day. There are drawings of cities as well, such as Ottawa's towers in the evening sun. An alder in autumn colors caught his attention on the trip from Montreal to Boston. The seventeen-day ocean voyage from New York to Genoa yielded seascapes, African and European coastlines, and the bizarre rock formations of the Mediterranean islands. These were quick sketches, to be completed at leisure in watercolor.[321]

Braun used the pictures to illustrate a talk he gave for several of his colleagues in his Strasbourg apartment. He prepared two exhibits for the occasion. One was a model of Old Faithful that spouted soda water.[320] The other was a model of an American shoe factory with an assembly line. When a handle was turned, little pieces of leather marched into one end of the model, and out the other end came little finished shoes.

A talent for drawing and painting ran in the Braun family. When he was young, Wunibald wanted to be a painter; and during the years when his brother Philipp was a principal in Hanau, he and Ferdinand often took their sketch pads and colors along on their walks through the Black Forest and the Vosges Mountains.[321] There are many pleasing watercolors extant from these cross-country trips.

In 1897 Johann Ambrosius Barth, publishers of the *Annalen der Physik und Chemie,* brought out a special volume (series 3, volume 63) to celebrate the 50th anniversary of the doctorate of Gustav Wiedemann, who had edited the journal for twenty years. Most German physical scientists of note were represented in it. Braun's contribution, "On Motions Generated by the Electrical Current," bore the dateline "Victoria (Vancouver), 6 September 1897." In this paper, Braun reported that a thin membrane separating two solutions bulges in the one direction or the other according to which solution contains the positive pole of an electrical battery. Quick reversals of the polarity cause the membrane to execute "movements large enough to be perceived visually. . . . To what extent the experiments reported herein are applicable to biological processes I leave to the specialists. Doubtless they will find the experiments useful on occasion in their own observations."[58] In particular, he suggested that recent electrical experiments with bone marrow and with amoebas might be repeated in the light of his new observations.

Braun had made these observations in Tübingen back in 1894. When the request—almost a royal command—for a contribution to the special Wiedemann volume reached him, he had one on hand. The measurements had been made with a biological membrane taken from a pig and with an old Saussure hygrometer, in which the usual hair had been replaced by a strip of the membrane. This antiquated instrument happened to be available in the Tübingen storeroom and had proved to be very handy for this purpose.

A pleasant surprise awaited Braun on his return to Strasbourg on 15 October 1897. In a back issue of the *Annalen* he found the first report from another laboratory of an investigation in which his tube had been used. The author was Berhard Walter, an assistant in the State Laboratory of the Free City of Hamburg, which was under the direction of Prof. August Voller, one of the "opponents" against whom Braun had defended his thesis. It was entitled "On the Processes in an Induction Machine." On the basis of theoretical considerations, Walter had predicted certain wave phenomena and was able to identify them with "relative ease by means of the cathode-ray tube, recently described by Prof. F. Braun." In Walter's work the name *Braunsche Röhre* ("Braun tube") appeared for the first time.

A surprise of a different kind appeared in the *Annalen* for March 1898, where the French scientist Albert Hess published a *Reclamation* in which he took issue with Walter: the idea of using magnetic deflection of cathode rays for the determination of current waveforms, Hess wrote, could not be said to have originated with Professor Braun.[269] "I take the liberty of drawing your attention to my own description of this application of the cathode tube, which I published three years ago and which was presented to the Academy in Paris by Prof. Cornu." After a hurried check of the French Academy's proceedings the Strasbourgers concluded that Hess's claim was unjustified. His paper of 2 July 1894, "On an Application of Cathode Rays for the Study of Changing Magnetic Fields," lacked Braun's simple method of making the results visible: the phosphorescent screen. Instead, the end of Hess's proposed tube was to be covered by a metal plate containing a "Lenard window," a thin strip of aluminum foil laid across an airtight, narrow slit through which the cathode rays would pass to the outside. They could then enter in similar fashion another evacuated chamber, in which the curve would be registered on a moving photographic plate. The arrangement was described "only in general terms, since details follow readily."

In his reply, Braun pointed out that the details did not follow readily by any means, since such proposals frequently came to grief precisely because of wholly unexpected practical difficulties; and no experimental proof had been offered. "In fact," wrote Braun, "when one considers the absorption of the cathode rays in the two aluminum foils, their dispersion in air, the difficulty of handling the photographic plate in the evacuated box, and the general unwieldiness involved in exposing even a single plate, one would have to be dealing with serious questions indeed before one would decide to resolve them experimentally in this manner. On the other hand, I have *shown* that the apparatus I have described provides an extraordinarily convenient and direct method over a large frequency range and already has proved itself to be useful in several cases."[59]

That was the end of the only real claim ever made against Braun's priority. A tube that Tesla had demonstrated in London in 1891, which later became the basis of an attempted claim by third parties, merely demonstrated the deflection of cathode rays by magnetic fields, but no practical design was proposed. The device that "has already proved itself to be useful in several cases" was used by Braun himself in the spring of 1898 to determine the distribution, direction, and density of magnetic lines of force. He reported this work in an engineering journal, *Elektrotechnische Zeitschrift*.[66]

Other investigations utilizing the cathode-ray tube continued to be made in Strasbourg,[135] but even before Hess's claim the device had begun to lead a life of its own. It became indispensable wherever rapidly varying processes were to be investigated or monitored. The number of its applications has continued to grow ever since. Even during Braun's lifetime innumerable versions of the device were constructed, from the tiniest to an enormous instrument weighing three tons that was used in Cambridge to investigate high-voltage breakdown. For a long time the Technical University of Aachen was a world center of research on the tube. A team of seven scientists headed by Dr. W. Rogowski investigated every conceivable application and construction. An important step forward was made when the hot oxide-coated cathode invented by Arthur Wehnelt was introduced as a source of the cathode rays (by then understood to be electrons); another advance came when, in addition to magnetic deflection, electrostatic deflection by means of condenser-like deflection plates was incorporated inside the tube. The "magnetic lens" attributed to Wiechert resulted in sharper focusing and greater brightness of the image.

Yet these were all improvements on the original, not new inventions.[131] Braun's original concept was so simple and at the same time so complete that few modifications were necessary or desirable. Moreover, as in all great inventions, the range of its applications soon extended far beyond the field for which it was originally intended: measurement and observation. Within a few years two of Braun's assistants, Max Dieckmann and Gustav Glage, recognized the fundamental significance of the device for the realization of mankind's ancient dream to see at a distance. In 1906, they patented their "method for the transmission of letters and line drawings by means of the cathode-ray tube." Their chief, though, thought that this was a form of "hocus pocus" and wanted absolutely nothing to do with it. Dieckmann recalled later that "he was not exactly enthusiastic about having his cathode-ray tube used in television experiments. Seeing at a distance was not an altogether respectable subject for investigation in those days—it was regarded much as were perpetual-motion machines."[116] One wonders what Braun would have thought of some of the applications his tube has found since then—radar, image intensification of x rays, monitoring space vehicles, and close-ups of the surfaces of planets and of the far side of the moon!

In view of these many ramifications, one well may ask why Braun never patented the cathode-ray oscilloscope. Did he fail to appreciate its far-reaching importance? Did he hesitate at patenting it because Röntgen had not patented x rays? The answer lies in his point of view. Braun wanted to give an instrument to science that would facilitate research on rapidly varying processes. He wanted every scientist to be able to make use of it without hindrance. That is why Braun never patented his tube: he thought that to do so would negate his goal in many cases. That is why he published such a detailed and exact description of his tube in the *Annalen der Physik und Chemie* that any skilled person would be able to build his own tube.

In retrospect, Braun came to regard his American trip as a last vacation before the great undertaking that started immediately upon his return: the development of wireless telegraphy. His first introduction to it came by way of three solid citizens—the *rentier* Friedrich Niess, the architect Gustav Gümbel, and the merchant Albert Zobel[307]—who were in possession of a discovery that made wireless telegraphy through water possible. According to a document written in Braun's own hand that is still in Annostrasse 27 in Cologne, the scheme was the invention of an "electrical engineer." It is no longer possible to determine whether the electrician in question was in fact the aforementioned Friedrich Niess, although we do know that Niess lived in the vicinity of the city wall in whose moat the first experiments were made, and that he is invariably named in the surviving documents ahead of Gümbel and Zobel.

The three entrepreneurs had been put in touch with Braun almost accidentally.[86] The latter was engaged to elucidate the scientific principles on which the invention was based, since the results obtained with it would not be reproducible unless the fundamentals were understood.[307] The three men were introduced to him by the German chocolate manufacturer Ludwig Stollwerck of Cologne, whom they had approached for financial support. Stollwerck wanted to know whether the "Strasbourg wireless telegraphy through water" could really fulfill its sponsors' promise.

Mankind's efforts to overcome the obstacles of time and space by communications surely date back to prehistoric times. In the Stone Age drum beats and smoke signals were used to send messages. Optical telegraphy ("writing over a distance") by fire signals was highly developed in antiquity. The first electric telegraph of Samuel F. B. Morse was a successful realization of an idea also tried by the Göttingen scientists Gauss and Weber, who had attempted to use the magnetic effects of electrical currents for the transmission of signals in 1831. The Morse telegraph required a wire between transmitter and receiver that had to be strung across unexplored continents and on the bottom of the sea. Almost from the beginning of the electric telegraph, the notion of wireless telegraphy occupied some of the best minds of the age.[346]

Conduction through the earth or through water, induction (the generation of current in a conductor by another current flowing in a neighboring conductor), and—after Hertz's discovery—electromagnetic waves were all regarded as possibilities for wireless telegraphy. Elaborate experiments were conducted with induction telegraphy. Edison tried a method of communicating with moving trains by signals sent through wires laid along the tracks. The chief engineer of Britain's telegraph system, William Preece, attempted, with moderate success, to communicate with an off-shore island by stringing long wires, one along the coast and one on the shore of the island, and relying on one wire inducing currents in the other despite the substantial distance between them. The idea of conduction through earth and water went back at least as far as Morse's 1842 observation, when one of his cables broke in New York Harbor, that the signals continued for some time after the separation. That discovery must have been made again and again by telegraph engineers the world over.

In 1894 Preece's experiments came to the attention of the major German electrical manufacturer Allgemeine Electricitäts-Gesellschaft (AEG) in Berlin. AEG engineer Erich Rathenau was dispatched to England and participated in several of the induction experiments. He suspected that both induction and conduction phenomena were responsible for the results. On his return he initiated experiments at the Wannsee, a lake near Berlin, where he sent dc through the water and was able to perceive a "soft, rumbling noise" up to a distance of 4.5 kilometers.[264]

This result brought nearer the possibility of wireless communication through the ocean with ships, lightships, and lighthouses. Government interest picked up slightly, especially in the German Navy. In 1895 the German Post Office continued experiments under the supervision of Dr. Karl Strecker, a telegraph engineer and university teacher, but he found the proposed experiments "unwieldy and time consuming, and one is obliged to base one's conclusions on quite inadequate data." In his report, which appeared on 13 February 1896, Strecker also observed imaginatively that others may regard these imperfections as a challenge: it was "perhaps this very circumstance that might stimulate others to work on this problem, since many interesting questions remain unsolved."[265] It was this remark that led Ferdinand Braun to his own work on wireless telegraphy twenty months later.

Electromagnetic waves, which had been observed even before Hertz's experiments, were not systematically used for communications until after Hertz. Their high frequency put them so far above the audible range that reflectors "as large as a continent" would have to be used to launch waves at audible frequencies, as Hertz himself pointed out in answer to a query from a German engineer working in the Netherlands.[347] The problem of super-imposing detectable signals on high-frequency electromagnetic waves was not solved until after Hertz's death in 1894. Another problem was the detection of the weak signals produced by the Hertzian waves. Hertz used a spark gap as a detector, but that only worked in the immediate vicinity of the source; a more sensitive detector was needed that would respond to the weak electromagnetic waves at some distance from the transmitter.

A discovery made in 1890 by Édouard Branly in Paris played a crucial role in the solution of this problem. Branly found that the high-resistance metal filings in a glass tube abruptly changed to a low resistance when high-frequency electromagnetic waves impinged on the device. The assumption was that the cohesion of the particles was affected by the electromagnetic waves, and the device was named the *coherer* by Oliver Lodge. (Branly did not care for the name. He would have preferred *radioconducteur*— one of the first instances of the use of the word *radio* in the communications context.[348]) The coherer remained the only practical detector of radiotelegraphy's first decade and ultimately earned Branly election to the French Legion of Honor.

Actually, a number of other investigators had noted similar phenomena, including Braun.[68] In his Leipzig experiments in 1876, while investigating whether the rectifier effect could be traced to changes in the contact with the crystal, he had noted that "tubes filled with metal particles likewise show changes in resistance caused by the induction current. The change in resistance persists even with a constant current."[9] The coherer effect had been described as early as 1835, by Munk, and again, after Braun, by Hughes in 1879 and by Calzecchi-Onesti in 1884. Yet there were no priority claims because, as Braun himself said, "The discovery that the resistance changes constitute such an extraordinarily sensitive reaction to rapid electric oscillations belongs to Branly."[68]

In the spring of 1894 Lodge used Branly's coherer to transmit and receive Hertzian waves. An important innovation was to place the clapper of an electrical bell connected to the coherer in such a way that each time its ring signaled the presence of an electromagnetic wave, the clapper struck the coherer and shook the particles loose, thus making it available for further reception. In Russia, the physics instructor A. S. Popov at the naval base of Kronstadt used Lodge's arrangement for a storm indicator he had built for the Institute of Forestry. Popov's receiver was attached to a lightning rod and recorded the atmospheric discharges (*static,* in modern parlance) that abound in the vicinity of a storm area. Popov read a paper, "On the Relation of Metallic Powders to Electrical Oscillations," at a meeting of the Russian Physico-Chemical Society in the spring of 1895, which was described in a single paragraph in the society's journal.[349] A more complete description containing some new data appeared in January 1896 under the title "Apparatus for the Detection and Recording of Electrical Oscillations," which ended on a speculative note: "In conclusion I may express the hope that my apparatus, when further perfected, may be used for the transmission of signals over a distance with the help of rapid electric oscillations, as soon as a source of such oscillations possessing sufficient energy will be discovered."[350] (The Russians subsequently based the claim that Popov "invented radio" on these two publications, but the claim is not generally accepted.[351])

A "source of such oscillations possessing sufficient energy" was devised by young Guglielmo Marconi (figure 16), son of an Italian father and an Irish mother, who had audited some of the lectures of Prof. Augusto Righi at the University of Bologna. Righi was one of the first physicists to incorporate Hertz's results into a regular physics course. In the summer of 1894, Marconi decided to test whether Hertzian waves could be used for signaling.[283] He experimented during the winter of 1894–1895 in two attic rooms in his parents' villa at Pontecchio near Bologna. When spring came, he moved the experiments into the open and by the early summer of 1895 was able to receive signals at a distance of 800 meters from the transmitter. As others had before him, he hit upon the idea of connecting an elevated metal plate (that is, an antenna) to one end of his transmitter and grounding the other end, which tripled the range of his apparatus.

In a matter of months, Marconi had achieved a distance comparable to Preece's induction telegraph and Rathenau's conduction telegraph. He accomplished this result by attaching large metal

Figure 16
Guglielmo Marconi (1874–1937).

plates as antennas to his spark-gap transmitter and grounding one
side of both transmitter and receiver. But his fame as *the* inventor
of radiotelegraphy rests above all on the astonishing persistence
with which he fought for the realization of his idea. Braun expressed it as follows: "Even though Marconi at first essentially
used only means which were already known, that in no way
diminishes his achievement in having taken the entire enterprise
seriously and having pursued it with great energy until he reached
practical, useful results."[68] Also worthy of recognition is the fact
that Marconi was neither scientist nor engineer, but an amateur
who must have spent many hours figuring out things that would
have been obvious to anyone with scientific or technical training.

On 2 February 1896 Marconi and his mother left for Great
Britain, which was then the only country interested in radiotelegraphy. Along its foggy coast, cable connections to outlyng
islands and lighthouses often broke, so that nobody knew what
was happening at sea. Ships could sink and their entire crews

drown while rescue vessels waited nearby unaware of the looming disaster.

Preece's induction telegraphy could bridge the gap only in the case of the larger islands close by, where there was room for the long transmission and reception wires to be laid out parallel to the ground. Marconi's system required only one dimension, to be found at any lighthouse: height. That was Marconi's strong point, as he knew perfectly well. His other advantage lay in the connections of his British relatives.[266] With his mother's encouragement and the appropriate introductions, Marconi found his way to Preece, but not before he had safeguarded his invention by applying for a British patent.

The first demonstration took place on the rooftop of the General Post Office in London in August 1896. Preece was immediately convinced of the utility of Marconi's system and arranged for an open-air demonstration at Salisbury Plain, during which a distance of 2.8 kilometers was covered by means that obviously were vastly simpler than Preece's induction telegraph. At a public lecture in London's Toynbee Hall on the last day of 1896 Preece announced that the British Post Office would spare no expense to initiate thorough experiments with Marconi's invention. The high point of the evening came when the young inventor started an alarm clock in the hall by wireless. In Germany, *Elektrotechnische Zeitschrift* commented that in reading about the Toynbee Hall Lecture one "was tempted to think of tales from telegraphy's fabled past, when deception was rampant." But meanwhile British financiers were busy floating a stock issue of £100,000 (then $500,000).

By March 1897 Marconi had matched Preece's accomplishment of bridging a distance of 5.3 kilometers. On 17 May 1897 he established the superiority of radiotelegraphy by communicating over the 13-kilometer distance between Pennarth and Weston-super-Mare; and on 20 July 1897, over 15 kilometers near La Spezia on the Italian coast. At about the time Braun was preparing for his trip to America, The Wireless Telegraph and Signal Company Ltd. was being formed in London, with Marconi as one of its five directors.

Germany's industry might have been indifferent, but not its government. The young and ambitious Kaiser Wilhelm II promptly dispatched his personal consultant to Britain to find out what he

Figure 17
Adolf Slaby (1849–1913).

could. This advisor was Prof. Adolf Slaby (figure 17) of the Technical University of Berlin-Charlottenburg, first holder of a chair of electrical engineering that had been established in 1882 at the suggestion of Werner Siemens. Slaby had caught the Kaiser's attention during a lecture series arranged for high government officials.[184] From that time on he remained the emperor's advisor on science and technology. An official invitation from Preece was arranged for Slaby to attend the March 1897 demonstration at the Bristol Channel in England.[273] He came away most impressed. Marconi was furious at the presence of a potential competitor, but could do nothing; it was Preece's show.[352] Slaby, on his part, was full of admiration. "The generation of Hertzian waves," he wrote, "their propagation through space, the sensitivity of the electric eye—all that is said to have been known. Very true, but with these known means one got just as far as 50 meters and no farther."[353]

Once back in Berlin, Slaby renewed his effort to match these achievements. As the Kaiser's science and technology advisor he had almost unlimited access to manpower and materials.[237] On 7 October 1897, while Braun was sketching the coast of Sardinia from the deck of the ship on which he was returning from America, Slaby had suspended two 300-meter antenna wires from balloons, by means of which he bridged the unprecedented distance of 21 kilometers.

Braun's first introduction to wireless telegraphy was the previously mentioned conduction telegraph of Albert Zobel and company. The transmitter was housed in a small wooden hut from which wires led to copper plates submerged in the moat outside the walls of Strasbourg—the same arrangement that Rathenau and Strecker had used in Berlin. The receiver was located a short distance away and was likewise connected to a pair of submerged copper plates. It consisted of a telephone receiver through which signals from the transmitter could be perceived. The system worked—that much was obvious to Braun right away. What was surprising was that the signals were uncommonly strong, much stronger than might have been expected from Rathenau's description. Braun's explanation was that chopped dc had been used in the Berlin experiment,[264] whereas in Strasbourg, with its up-to-date generating station, ac had been available. Alternating current, especially when "harmonics" that are multiples of the fundamental frequency are present, tends to travel near the surface (*skin effect*) and thus can be picked up more easily than dc.[68] That was apparently why the Strasbourg apparatus yielded much stronger signals than the one in Berlin.

The history of radiotelegraphy in Germany is interwoven with that of Albert Zobel, a courageous businessman who is nowhere commemorated. Zobel was a Berliner living in Strasbourg. By 1887, when he was 31, he had accumulated a modest fortune that enabled him to live in an exclusive quarter of the city.[203] We do not know how or why Niess and Gümbel first approached him, but he became immediately interested in the invention they brought him. He could see that its further development would require money—lots of money. Zobel undertook to raise it. In the summer of 1897 he approached Ludwig Stollwerck, who was known to be interested in promising technological innovations.[268]

Ludwig Stollwerck was one of the five sons of the confectioner

Franz Stollwerck, who started a chocolate factory in Cologne in 1839 that afterward became world famous. The five sons divided the task of running the growing business among themselves. Ludwig specialized in marketing and diversification. Through his supplier of shipping materials, the firm of Metallwerke Theodor Bergmann in Baden, Ludwig Stollwerck was introduced in the mid-1880s to Max Sielaff, the Berlin inventor of a vending machine whose use in the merchandising of chocolates became one of the success stories of the century.[268] Through his interest in vending machines, Ludwig Stollwerck made other useful contacts. He became interested in the phonograph and got in touch with Thomas Alva Edison in New York, encouraging him to turn his invention into a dictating machine; he asked the inventors of the *cinematograph,* the brothers Lumière in Lyons, to give a demonstration in his plant—the first film showing in Germany;[268] and he became interested in electrical engineering, which led to his interest in wireless telegraphy.

Although Ludwig Stollwerck was deeply impressed by the Strasbourg invention, he subjected it to sober scrutiny. He asked his London branch, Stollwerck Brothers, to determine what The Wireless Telegraph and Signal Co. had paid Marconi for his invention, so that Stollwerck could get an idea of its economic worth.[307] He also asked a scientific acquaintance in Berlin, the physicist J. Wrede, to determine the invention's technical value. But Wrede sent his "great regrets that he was not sufficiently well informed on recent electrotechnics to give a professional judgment." Instead he recommended Slaby; "otherwise my friends among the engineers at Siemens & Halske or at AEG."[307]

Neither alternative pleased Stollwerck. A test by Slaby would have delivered the invention to an acknowledged competitor. A test by the engineers of one of the major companies would have exposed it to potential competitors. Stollwerck and Zobel finally settled on Braun, who was known to be an experienced "electrician."

Braun's belief that the skin effect came into play when ac was used had led him to the conclusion that even greater efficiency could be obtained at still higher frequencies. On his part, Stollwerck came to the conclusion that further development of an invention based on such complex scientific principles could not be left to the scientifically naive original inventors and he made

Braun's participation in the venture a condition for his own. Not until 8 February 1898 did he receive the long-awaited telegram from Zobel: DESIRED AGREEMENT REACHED LAST NIGHT WITH THE PROFESSOR.[307]

Now there was money for new experiments. Braun designed transmitter and receiver transfigurations, and Niess and Gümbel tried them out, first in the old moat and then in the Rhine, in its tributary the Kinzig River, and in the gravel along the riverbanks. Braun's designs of the winter of 1897–1898 contained all the basic points of the work for which he later received the Nobel Prize:

1. In his search for a high-frequency oscillator, Braun first tried Hertz's configuration and found it unsatisfactory. Braun concluded that waves of such high frequencies, known to propagate in a straight line, could not "turn the corners" and follow the course of the waterways along which the experiments were being made.[68] He turned instead to the lower-frequency "Leyden jar discharge" that Feddersen had used in 1858.[354]
2. In this scheme, a Leyden jar was discharged through a coil, providing oscillations that persisted for a short time, at a frequency that could be lowered to a more manageable range simply by an enlargement of the coil (figure 18). The combination of jar (capacitance) and coil (inductance) constituted an oscillating circuit.
3. Braun knew from theoretical considerations that to make the oscillations persist, the circuit would have to be as free from external "damping" as possible, just as a pendulum or a weight on a spring must be as free as possible from friction if its oscillations are to persist. Connecting his transmitter through wires to plates submerged in water represented severe electrical damping. It would be better to couple the transmitter to the plates via a pair of coils, a primary coil connected to the transmitter and a secondary coil, to the wires; then the secondary circuit could remove energy from the primary circuit without damping it severely.
4. Finally, Braun determined that the results could be significantly improved by an adjustment of the number of windings in the primary and secondary circuits; that is, the arrangement worked best when the two circuits were in resonance.

These four characteristics—increase in transmitter power, use of an oscillating circuit, minimization of damping by use of in-

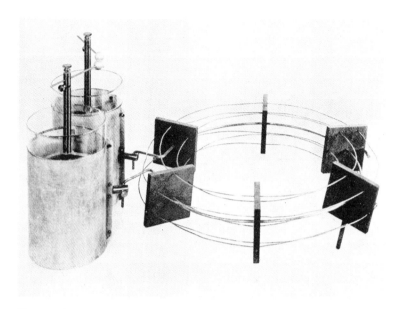

Figure 18
Braun's oscillator circuit in a version comprising coil (right) and two
Leyden jars.

ductive coupling, and resonant "tuning"—were the distinguishing
features of Braun's system of radiotelegraphy. Even today transmitters the world over work according to these principles.

The Strasbourg Physics Institute's massive stone façade presented a calm appearance to the outside world in the spring of
1898. The bustle of radiotelegraphic experiments that were to
begin a few months later had not yet started. Yet the external
calm was belied by the intensive activity inside. Braun's lectures
in experimental physics (sound, heat, and electricity in the winter;
mechanics, molecular physics, and optics in the summer) always
attracted at least 200 students. Laboratory exercises were attended
by dozens of students. Braun went from bench to bench, encouraging the students and offering each a friendly word. To a
student using a sodium light for an optics experiment: "Watch out
you don't burn down the optical axis!" Polite laughter all around.
To another, working with a barometer: "What does the barometer
read when it falls?" The student has no idea and is directed to

perform the experiment. The result: the pressure reading of a barometer in free fall is zero.[325]

There were always at least half a dozen scientific projects under way. The experimenters all came to Braun for advice and help. He also had to prepare the now traditional colloquium, do the administrative work, participate in examinations and faculty meetings, and receive an unending stream of visitors. Two councilmen from the Alsatian town of Hagenau complained that their new bell did not sound right. Zenneck was dispatched to check it and found that the clapper was hitting the bell in the wrong spot because the raised inscription had disturbed the original axial symmetry; rehanging cured the trouble.[135] A mother came to ask whether she should let her son continue to study. "My dear lady," said Braun with a twinkle, "not everyone needs to study. I believe your son would do just as well in life if he became a good shoemaker!"[321] And then there were introductory visits by new students, new colleagues to be shown around, and so forth.

Despite all that and his growing interest in radiotelegraphy, Braun did not neglect his own scientific work. Four little essays appeared in the *Annalen* in May 1898. The "Note on Thermophony" described a method for converting the current fluctuations of a telephone transmitter and reconverting them back into current fluctuations farther along the flame by means of a wire grid.[60] "On the Emission of Light from Various Electrolytic Electrodes" dealt with observations that Braun had made in the course of his investigations of electrolysis through a narrow slit. In addition to gas bubbles, he also had discerned tiny sparks.[61] "A Criterion for Determining the Cohesion of a Surface Layer and a Note on the Vapor Pressure of Such Layers" reported on his scientific correspondence with Vladimir Shchegleov (who had made Braun's acquaintance during Braun's first stay in Strasbourg and in 1886 was appointed professor at the Technical University of Moscow) in Moscow about the 1896 work with plaster disks.[62] "Do Cathode Rays Exhibit Unipolar Rotation?" examined the behavior of an electron beam under the influence of an axial magnetic field.[63] The surprising result: "Instead of a rotating figure, a stationary cone-like configuration is obtained," which is shown on the phosphorescent screen as a ring of light—a finding that Braun had mentioned in Toronto.[58a]

Once the conduction telegraphy experiments had achieved a range of 1,600 meters in water and 800 meters in earth, the method was patented, on 12 July 1898.[64] The Berlin patent attorney, Carl Pieper, listed the following claims:

no perturbing, space–consuming, or dangerous wires (of the sort used in induction telegraphy);

signal strength does not decrease as the inverse square of distance (as in wave telegraphy); and

topographical obstacles cause no interference (as in Hertzian short-wave telegraphy).

The patent application was signed by Niess and Gümbel but Braun's name appeared as well, in recognition of the increasingly active part he had played. Stollwerck received a wire telling him that the patent application had been submitted. On the telegram form, he noted with satisfaction, "Nothing can stop Professor Braun now."[307] He was quite right.

Why did the group apply for a patent on a method of "wireless telegraphy through water and earth" with a range of 1.6 kilometers at a time when Slaby, using the Marconi system, had already achieved a range of 21 kilometers? There were several reasons. In 1898 only one method seemed feasible, based on the Hertz configuration. But that was blocked by Marconi's patent. Anyone else who wanted to work on wireless telegraphy would necessarily have to restrict himself to other areas. Water telegraphy was such an area.

Another consideration was the lively publicity the Marconi system was getting with extravagant reports on his string of successes, generally without any technical details. The boxes housing Marconi's apparatus always remained locked.[250] Marconi and his company spoke only reluctantly about the details of his invention, allegedly to protect the patent, but at least in part because neither Marconi nor any of the scientists who had participated in his experiments were too clear in their minds about the fundamental phenomena.

In contrast, the "conventional" methods of wireless telegraphy were well understood. Braun could describe the principles on which his water telegraphy was based quite precisely and say what conclusions could be drawn from them, whereas no theory existed to explain the results obtained with Marconi's system and there was no way to prognosticate its further development.[121] According to Marconi's empirical formula relating range and height of the antenna, Slaby should have achieved a range of 150 kilometers in 1897, but in fact had attained only 21 kilometers. "During one of Marconi's trials," reports one account, "he spent two fruitless days trying to get his apparatus to work. Then he added twenty yards of wire to the antenna and signals were received perfectly."[121] No one could say why. When in 1898 the British military authorities established a wireless connection between Fort Lavernock near Cardiff and Fort Flat Holm, they used not Marconi's system but the more reliable Preece induction telegraph.[250] If an alternative to Marconi's system was considered viable in Britain itself, authorities elsewhere certainly could not afford to dismiss other alternatives, at least not in the summer of 1898.

As soon as Braun's patent application had been submitted, Stollwerck tried to interest British financiers in the Strasbourg water-telegraphy system. His plan was to apply for a British patent as soon as the system had matched Marconi's maximum range to date, 50 kilometers, and to establish a British subsidiary that would compete with Marconi's company. At least one British industrialist was ready to cooperate: the soap manufacturer William Lever in Port Sunlight near Liverpool. Stollwerck had a business connection with him through his interest in vending machines, which could be as readily adapted to dispense cakes of soap as bars of chocolate.[307] But another business acquaintance of Stollwerck's, Smith of the Edison-Bell Co. in Glasgow, poured cold water on the scheme. "I feel a great deal of work remains to be done," he wrote Stollwerck, "before this invention can be of practical use. Even if it were perfect I am not at all sure that it would be of more value than Mr. Nansen's North Pole expedition."

That ended Stollwerck's bold "English project." Smith's sarcastic rejection made him wonder whether he had let himself in for a hopeless adventure. Another blow fell on 6 September 1898, when the German Patent Office rejected Braun's application because of Strecker's previous experiments. Braun wrote the Patent Office and Stollwerck immediately, explaining that unlike Strecker, who had employed quasi-stationary chopped dc, Braun had used ac, which alone was capable of surface propagation. Stollwerck nevertheless asked his friend Franz Joly, the director of the Cologne electric company, and Professor Slaby of Berlin to go to Strasbourg and check out Braun's system. Braun promptly resigned, writing to Stollwerck that he would have welcomed the opportunity to discuss his experiments with Slaby as a friend but that he could not accept him as "your professional evaluator."

It was shortly before this crisis that Braun, who had naturally made it his business to learn all he could about competing systems, had arrived at a conclusion that was to be of the greatest importance in the development of radiotelegraphy. Although Marconi had gone from success to success, for instance receiving worldwide coverage of his feat of reporting the Royal Regatta on 4 August 1898 by wireless, he had actually made no technical progress during the year just past. To exceed a range of 15 kilometers a disproportionately large amount of energy became necessary. Braun asked himself why it was so difficult to increase the range. If everything functioned properly at a distance of, say, 15 kilometers, why could not twice or several times that distance be achieved simply by increasing the potential, something that could be done quite readily?[96] Was there some inherent limitation?

The answer dawned on Braun during the summer of 1898, when it occurred to him that Marconi's design, with the spark gap inserted between antenna and ground, was reminiscent of one of the earlier connections for the water telegraph—the connection that had been the *least* satisfactory![68] With that as a starting point, Braun quickly concluded that if that circuit had proved so unsatisfactory for conduction telegraphy, it was very likely the factor limiting the range of Marconi's system.

It took but a few hours to test this conclusion. Braun set up Marconi's circuit, but with the modification that had proved so effective in improving the conduction telegraph: a primary coil in the oscillating circuit and a loosely coupled secondary coil to transfer the oscillations into the antenna-to-ground circuit. The experiment was a success of decisive importance. Thereafter the Strasbourg conduction telegraphy would be only of historical significance (figures 19–21).

It was just after this that Braun received the announcement of a forthcoming visit by Adolf Slaby. Despite having witnessed

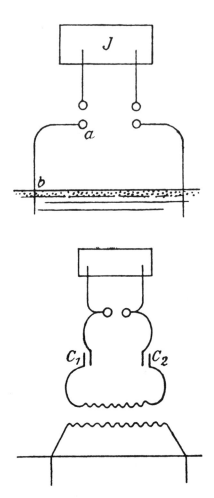

Figure 19
Top, poorly functioning transmitter for "water telegraphy" (February 1898); bottom, properly functioning version (May 1898); after Braun's drawings, rotated through 90°.

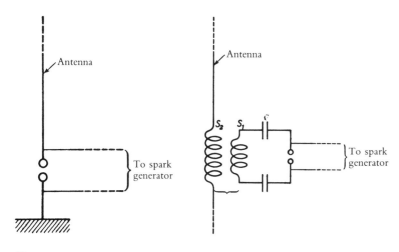

Figure 20
Left, Marconi's transmitter (1895–1901); right, Braun's transmitter
(September 1898); after Jonathan Zenneck's drawings.

Marconi's demonstration, despite having worked on various versions of his circuit for over a year, Slaby had not recognized, much less been able to eliminate, one of its major shortcomings. And now he was to be the expert whom the rightly concerned Stollwerck expected to evaluate Braun's apparatus! No wonder Braun had resigned. Yet Stollwerck did not want to lose his entire investment. He asked Zobel to see what he could do to bring Braun back to the fold. These efforts were successful. Slaby's visit was canceled and Braun resumed his work. His first step was to invite Stollwerck and his private evaluator Joly to Strasbourg on 20 September 1898 for a demonstration of his new wireless telegraphy.

Switching to aerial telegraphy had been such an abrupt decision that the demonstration had to be largely improvised. The transmitting antenna was strung between the laboratory and the institute tower. The receiving antenna, about 1 kilometer away, was another wire attached to a hops vine pole. "Across a line of sight, with the coherer adjusted to maximum sensitivity, the Marconi circuit gave barely perceptible signals," reported Braun. "With my circuit the signal strength immediately increased, even though the source was the same. Even when the receiving station was

Figure 21
Coupling coil used in one of Braun's early transmitters.

1898

shifted so that a church blocked the transmitter, casting an electrical shadow, the signal strength with my arrangement remained unchanged; the signals were perceptible even when the receiving pole was placed a few meters behind an equally tall bush in full leaf."[75] The participants' enthusiasm was unbounded. "We have something here better than anything we are ever likely to get," was Zobel's exuberant judgment.[307]

The commercial significance of Braun's invention was that it effectively broke Marconi's monopoly in radiotelegraphy. Braun's "new circuit," as it was first called, made a system independent of Marconi's possible, a fact of tremendous importance quite aside from its technical superiority. Application for a patent was made immediately.[65]

Subsequent experiments brought further rapid progress. On the day the patent was submitted, 14 October 1898, Zobel wrote to Cologne that "Braun has proved scientifically that he can broadcast three times as far as Marconi." Braun, however, refused to make any premature predictions. "What a cautious man!" wrote Zobel.[307] But on 25 October 1898, Braun declared that his circuit would bridge a distance of 100 kilometers and that he had made "such progress today that he thought it more desirable to own stock [in the company that was to be formed to exploit the invention] than cash."[307]

To appreciate what Braun's enthusiasm meant to the partners one must recall that Marconi, who had remained at a 15-kilometer range throughout the summer of 1898, could in the coming year nevertheless count on a doubling in the investments in the company that bore his name. At the same time, Braun's express wish for more shares in lieu of cash payments did not shake Zobel's determination to stick to the "previously concluded agreements," as he reported to Cologne.[307] According to these agreements, a telegraph company with a capital of 333,000 marks was to be founded on November 15 in Strasbourg or Cologne. Braun, Niess, and Gümbel as the inventors were to get shares in the amount of 37,000 marks each, Zobel for "finding the capital" was to receive 27,000 marks, and a group of five businessmen—Stollwerck and William Lever, Josef Schmitz-Schlagloth, Josef Wehner, and Albert Gehner—was to bring in 95,000 marks.[307]

In Marconi's original circuit, with the spark gap placed between antenna and ground, the same circuit that generated the oscillations also radiated them; the inherent limitation (which was soon reached) was that each function adversely affected the other. By separating the two functions, so that the oscillations were generated in a closed circuit that was coupled to a separate antenna circuit, Braun could be said to have not merely improved Marconi's system but introduced an entirely new principle, on which the future of radiotelegraphy was to be based. The "Braun transmitter" thus might be regarded as an invention of the same rank as the "Braun tube," and it has been described in much the same terms:

The usual characteristic of a truly basic invention is that its significance extends far beyond its originally intended application. That is particularly true of the closed oscillating circuit introduced by Braun. . . . Since Braun introduced the closed circuit, transmitter technology has undergone many changes. But even the most modern apparatus utilizes the closed circuit as a means for achieving maximum efficiency.[123]

In addition to extending the range of radiotelegraphy, Braun's circuit had the advantage that the decoupled antenna would no longer have to be at high potential. Thus it was not only much safer but also less susceptible to accidental short circuits.

That was not the case in still another method of radiotelegraphy that made its debut in 1898, that of Nikola Tesla, the inventor of the ac generator and of an air-core transformer called the Tesla coil. Without any evident plan, Tesla had adapted his coil to the transmission of energy at truly gigantic dimensions, with generators that produced millions of volts. It was not clear what Tesla's aim was as his work was conducted in secret.[241] He hinted that he was occupied on "wireless" work, the wireless transmission of electrical power, and "death rays."

Among other things, Tesla had contrived to steer a boat by radio, an experiment that might be regarded as the forerunner of missile guidance.[355] He was now reported to have announced that he would guide a boat at the forthcoming Paris World Fair by remote control from New York. When Braun was shown the newspaper clipping, reported Zobel, "Braun laughed about Tesla: 'Utopias and illusions, great fantasies. He has accomplished nothing since his discovery of the Tesla currents.' "[307] Yet four months later Tesla caused consternation in Strasbourg. A garbled cable from Stollwerck's New York representative, dated 8 November 1898, reported that a patent had been issued to Tesla for some sort of invention.

Since the founding of the Strasbourg wireless telegraph company was imminent, Stollwerck had the entire newspaper article cabled across the Atlantic. In great excitement he wired Zobel, "Tesla even claims he can steer ships at any distance by radio. At worst we must assume that Tesla has made the same invention as Prof. Braun. In that case the first 60,000 marks are lost." The wires flew between Cologne, Strasbourg, and Berlin. The partners even attempted to confer by telephone, but what with the primitive systems of the day and the speakers' great excitement, such communications proved most unsatisfactory, and the difficult decision had to be worked out by telegram. It was to postpone the founding of the company by one month.

Braun remained calm throughout the storm, but on 13 December 1898, two days before the month had run out, even he lost his nerve. He had received a reprint of Tesla's US Patent No. 645,576, "System of Transmission of Electrical Energy,"[275] and warned Stollwerck, "I consider Tesla's patent of *grave importance for us*. . . . We have lost a great deal of ground." To save what could be saved, he reverted to his original suggestion that the company's founding should go ahead, but that in view of the clouded patent situation his name should be kept out of it until a range of 100 kilometers had been attained.

The company was founded on 15 December 1898 as planned. One of the prospective participants, Albert Gehner, had withdrawn. Braun was also absent, his rights safeguarded by a secret agreement with the new company, whose name was to be Funkentelegraphie GmbH Köln. Braun took an active interest in the happenings at Cologne. He sent Niess several messages.[307] In one of them he asked that, after the incorporation, a note "about the results of our experiments" be sent to the *Kölnische Zeitung*.

But it was borne in upon him that he was no longer in control of the company's affairs. This particular request was not executed. Yet Braun considered such publicity "very important, to salvage our *factual* priority and thus to exert a benign pressure on the Patent Office."[307] The new board of directors heeded their attorney's advice instead—which was to say nothing in public until he gave the sign. In effect, Braun was muzzled against his better judgment, with dire results. Unlike the publicity-happy Marconi concern and the scarcely less voluble Prof. Slaby in Berlin, Braun made no announcements about his discovery until twenty-three months later, a circumstance that was to trouble him for years.

The Tesla threat disappeared toward the end of 1898. Even though the drawing of his apparatus showed an astonishing similarity to Braun's transmitter, the intent was quite different. Tesla wanted to use the upper atmosphere as a gigantic vacuum tube for the distribution of cathode rays.[275] His main purpose was the transmission of electrical power. As was his custom, though, he had introduced, in attorney Carl Pieper's words, "all possible ideas into the patent description without having had any definitely successful experiments to corroborate them." Here he had mentioned that his invention in principle could be used for the long-distance transmission of messages; and this was the idea that had alarmed Braun.[307]

Tesla's system had to be considered utopian, if only because of the huge technical investment it would have required. "Once again it has been shown that Tesla's publications are not worth the wind by which they are dispersed," wrote Pieper from Berlin, "and what's more, there is no danger that Tesla, by himself, is capable of a profundity of the kind that Professor Braun has demonstrated in his thoughts and actions."

On 27 January 1899, Braun delivered the traditional Emperor's Birthday Address to a packed auditorium at the university. His topic was "Methods of Research in Physics."[67] In this one-hour lecture, Braun sketched the history of physics from Aristotle down to the electrical-power transmission line between Lauffen and Frankfurt and offered some thoughts on possible future developments. The mature scientist—he was now 48—was capping his somewhat stormy career by making his peace with "philosophy, and particularly metaphysics, whose fundamental questions keep coming back with elemental force." By continually "reconciling existing knowledge with the unending search for the unknown," one might hope to provide a basis for a fruitful coexistence of the scientific and philosophical disciplines.

It was his encounter with the practical world that led Braun to make a very progressive proposal, "first bruited about among some thoughtful observers here, both inside and outside the university, to supplement pure research at our university with applied research, perhaps as a sixth faculty [that is, in addition to theology, law, medicine, letters, and science]." He saw that many would oppose the idea, which was in fact many decades ahead of its time; he nevertheless urged this "transformation of the *universitas* to enable it to participate autonomously in the great movements to come in the twentieth century."

In the fall of 1898 Braun wrote Ludwig Stollwerck that "from all the newspaper accounts I have read, Marconi still seems to be stuck as fast as ever. I am all the more inclined to compare his arrangements with ours on the high seas." Stollwerck consulted a Hamburg business associate, Georg Wilhelm Bargmann of the Hamburg-Manila Line. Bargmann pointed out that the mouth of the Elbe River offered a most suitable area for experimentation. Signals and lighthouses and a series of lightships out to the island of Helgoland made it possible to test wireless telegraphy at ever increasing distances. Since this was one of the main shipping lanes, the attention of the shipping industry would be focused on this novel enterprise. Moreover, in Cuxhaven the Stollwerck group would be on friendly territory: since 1394 the Free City of Hamburg had controlled this region, and Hamburg owed a considerable debt to the House of Stollwerck, which since 1860 had made it the leading cocoa market of Europe.

The first step was a visit by Braun and Zobel to Cuxhaven's representative in the government of Hamburg, Senator Max Predöhl. After half an hour's conversation, the senator placed "all buoys, lighthouses, and lightships in the Elbe estuary" at the disposal of the company for its experiments and directed the Cuxhaven authorities to extend all possible courtesies to their guests.[308]

The next morning Braun, Zobel, Bargmann, and Dr. Cantor (Braun's assistant from Strasbourg, who had been engaged as "technical director") took the train to Cuxhaven. They engaged rooms in the Hotel Dölle and began their preparations. A local mechanic, Heinrich Gock, and his helper Wilhelm Schwenk were hired to erect antennas for testing the apparatus at the Alte Liebe lighthouse and at a nearby pavilion "88 paces away." Visits were made to the coast guard and to local officials to offer assurances that no harm would come to people who might accidentally touch the antennas and that the experiments would have no effect on the magnetic compasses of passing ships.

After two weeks of preparations, the first experiments took place in April 1899. Braun, though, already had returned to his professorial duties, leaving Cantor in charge at Cuxhaven. The experiments made excellent progress (figures 22–24). At the beginning, the range was no greater than the distance to which the noise of the sparks themselves could be heard.[176] By May signal flags and a telescope had to be brought up to provide independent communications between transmitter and receiver, now 3 kilometers apart. By June the signal flags had to be run up a tall mast; the receiver was now on the island of Neuwerk, 12 kilometers away!

Figure 22
Kugelbake transmitter after spring flood: operators' shack containing instruments has been raised 5 meters and anchored in framework as protection against flooding.

Commercial development was equally rapid. Funkentelegraphie GmbH Köln, with a capitalization of 333,000 marks, had come to the end of its usefulness after only a few months. Zobel and Stollwerck convinced Braun that further development would require a much broader base. For that it would be essential to let the world know of Braun's participation, even though the agreed-upon 100 kilometers had not yet been reached. The result was the formation of Prof. Brauns Telegraphie GmbH, capitalized at 700,000 marks, with Ferdinand Braun at the head of the "inventors' group." Yet even this company was to be transitory. Its main purpose was to enable Zobel and Bargmann to raise even greater sums.

On 24 June 1899, the partners invited a group of representatives of Hamburg's financial community to Cuxhaven. While the wealthy shipowners watched, "messages were delivered by Prof. Braun's telegraphy from the lighthouse to a buoy 3 kilometers away. They were processed by the receiver as with ordinary tele-

Figure 23
Transmitter adjacent to Alte Liebe lighthouse, 1899, one of the earliest photographs of wireless-telegraphy equipment in field tests.

Figure 24
Dr. A. Köpsel, Ferdinand Braun, Jonathan Zenneck (left to right) in
Helgoland wireless station.

graph."[277] On the excursion steamer by which the party returned
to Hamburg, there was a festive dinner at which Hamburg's Lord
Mayor, J. G. Mönckeberg, toasted the group on their outstanding
achievements. Two weeks later, on 7 July 1899, Bargmann, who
was related to the mayor, was able to announce the formation of
a syndicate for the commercial exploitation of Braun's invention.
Albert Zobel was president, Bargmann secretary, and Stollwerck
chairman of the board.[308] Capitalization was 2,000,000 marks, of
which the Hamburg merchants, shipowners, and attorneys had
provided 300,000 marks in cash.

The formation of the syndicate, which soon became known by
its telegraphic address, Telebraun, marked the end of the "Cologne
period" of wireless telegraphy. Although Stollwerck was still the
largest stockholder and chairman of the board, Hamburg was now
its economic center. On 13 July 1899, the Hamburg Senator Hol-
thusen headed another party of visitors to Cuxhaven.[277] And the
following week, while passing the installation on his way to Hel-

goland, the Kaiser noticed Braun's lighthouse mast and learned for the first time that Slaby was not the only person in Germany experimenting with radiotelegraphy.

Meanwhile, back in Strasbourg, another kind of visitor had found his way to Braun: a representative of the world's largest marine insurance company, Lloyd's. Lloyd's owned coastal signal stations all over the world to communicate visually with passing ships. A contract to equip these stations with radiotelegraphy would have been the largest order in the world. Marconi's Wireless Telegraph and Signal Co. was quite aware of that and had already submitted a quotation. But the management of Lloyd's had learned that a university professor in Strasbourg had devised "a new circuit." They decided to find out for themselves which system was better suited to their purpose. The Lloyd's representative, Hozier, invited Braun to go to England and conduct experiments there, where the company would put "all facilities" at his disposal.[274] But nothing came of it. Lloyd's decided that the requirements of Braun's system, which called for the primary and secondary circuits to be tuned to each other, made it less advantageous than Marconi's "simpler" untuned circuit!

Braun's external environment had changed radically in a matter of a few months. The quiet scholar had been torn from the relative tranquility of a university and plunged almost overnight into a storm center of worldwide commercial and political interests. Yet by all accounts he remained, what he had always been, a straightforward and unpretentious scientist.[132] As a matter of fact, his work in collaboration with industry meant many inconveniences for him. He sometimes longed to be back in his laboratory doing pure research.[131]

Braun's personal life remained largely as it had always been. The same man who had just concluded a negotiation with the world's largest insurance company and whose name was on the tongue of high government officials or board members of international cable companies (which regarded the wireless newcomers with misgivings) could still be seen taking his four children for a walk on a Sunday: Siegfried, now 13, Hildegard, 11, Konrad, 8, and Erika, 6. These walks were not merely for Ferdinand Braun's recreation, though. For many years, his wife had been ailing, and Braun had attempted to find substitutes for her in various ways, so that the children could enjoy a carefree youth. In bringing up his sons, Braun tended toward a practical education. As soon as they were old enough, they found a toolbox containing a hammer, a saw, and electric wire under their Christmas tree.[320] Konrad received special praise for putting together a galvanometer from these simple means plus a sewing needle. In the years that followed, his equipment came to include a battery-operated motor, a spark coil, Geissler tubes, and a wet battery. Konrad was quite at home in the institute workshop—supervised by Julius Rolf after the autumn of 1899—and in the large auditorium, after the day's classes.

On one occasion Konrad accidentally nicked the edge of a glass belljar used in the demonstration experiment in which an electric bell becomes inaudible once air has been pumped out of the belljar. Konrad attempted to repair the damage with glue and even made sure that all was again in working order. But during the acoustics lecture next morning the bell would not grow silent no matter how long the mechanic kept pumping. The glue had come loose overnight. Traces of it led to the "criminal's" trail. At lunch Braun asked his younger son whether he had been in the lecture hall the previous evening. When Konrad admitted that he had been there, his father said simply, "Next time you have a mishap like that with the belljar, say so and don't let the mechanic pump away for a quarter of an hour."[326] That characterized Braun's mild yet firm way of bringing up his children.

The sea is an unruly partner. Its whims prevented Telebraun from reaching its original goal of 100 kilometers by the winter of 1899–1900. As summer came to an end, it became clear that the period of experimentation at Cuxhaven would have to be extended.

The Strasbourg landlubbers had had a beautiful plan based on their prior experience in a well-equipped institute where experimental conditions could be changed easily and quickly. Now they found that "many time-consuming and deadening obstacles were being encountered. The sea wind exerted unexpected pressure against even single wires; it could take hours before the effects of small experimental changes, such as the insertion of a coil, were known; at the crucial moment the wind tore down guy wires, sudden fog made optical communications impossible; . . . and besides, the station on the island of Neuwerk was accessible only at low tide."[68]

The first attempts at transmission from the Neuwerk lighthouse failed. The copper facing of the lighthouse was thought to be at fault, so two masts, each 45 meters high, were ordered for installation on the breakwater. With the strong wind, though, the carpenter could manage only 35 meters. When contact was finally established, a wholly unexpected spring tide abruptly ended the experiment. The masts were knocked over and carried away by the flood. It was all the crew could do to save the equipment.[276] The sea seemed to have its own plans. Expenses mounted. In addition to local laborers, two special mechanics had to be brought in and a workshop improvised in which the day-to-day modifications could be made. Cantor insisted on the most modern equipment for the workshop. To charge the transmission batteries, for example, he brought in the first gasoline engine ever seen in Cuxhaven, a Daimler machine with spark ignition. It was very temperamental. "Some days it would not start at all; another time it started right away, nor could we ever figure out why it chose to behave that way. . . . It continually gave us anxiety."[135]

"One day," reported Schwenk, "Herr Cantor appeared in the workshop and demanded the immediate construction of a metal tower 20 to 30 meters high. Money was no object. We worked all night, and in the morning the costly tower was delivered. It was put on a glass plate near the lighthouse, but after half an hour it was discarded. The experiment had failed." These were difficult, adventurous times. Cantor might have gone on with it, but his style of work was somewhat too uneven. "He might worry for a long time about how to realize a plan, but if the result was not up to his expectations, he sometimes simply abandoned the work."[135] That was doubtless why Cantor did not extend his six-month contract with the company and returned to Strasbourg in October. However, he had received Telebraun shares worth 15,000 marks for having found the optimal tuning of the Cuxhaven apparatus; that is, he had established the best arrangement of coils necessary for resonance between the oscillating circuit and the antenna circuit.[307]

The North Sea gave him a spectacular sendoff. On September 21 a high tide, whipped by a storm, tore down the radio cabin on the buoy and carried it and all the apparatus and tables out to sea. It knocked over the great transmission mast, which only the week before had been strongly tied down for the winter. The fallen mast stove in the roof of the main transmitter station, which had also been recently strengthened. So had the outlying station, which had already shifted by a few meters in the above-mentioned spring tide—an incident that had caused 1,200 marks' worth of damage.[68] After the storm flood, the damage ran into the tens of thousands of marks.

Jonathan Zenneck replaced Cantor in Cuxhaven in October 1899, just after the first ship-to-shore experiments had taken place.[135] (A commemorative tablet, not entirely accurate, marks the spot.) For its experiments from a moving ship to land, Telebraun had acquired the use of the excursion steamer *Silvana,* which in summer carried vacationers to Helgoland. While the steamer was being made speedier and more stable for more strenuous winter duty in September 1899, Cantor and Schwenk crammed its lounge with apparatus. Twice weekly, on the round trip to Helgoland, various oscillator circuits and antenna combinations were tried out, with progressively better results. By December 1899 a distance of 35 kilometers had been achieved.

Zenneck made his first trip to Helgoland the day after his arrival. He was horribly seasick but valiantly took care of the equipment until the island came into sight. He later found his sea legs and was able to report that he could eat baked beans and bacon in a gale aboard an anchored lightship "whose motion is such that one might wonder whether the six degrees of freedom assigned by the laws of mechanics to a rigid body are sufficient for such a ship."

He had reason to remember two later trips on the *Silvana.* During a particularly heavy roll the transmitter coils, normally submerged in oil for insulation, partially emerged from the liquid and arced over with a gigantic burst of flame and set fire to the container's wooden walls. But on the next roll the ship leaned to the other side, the coil was resubmerged, and the oil extinguished the flame.[69] When the owners found out about this incident they naturally became concerned and threatened to withdraw their ship if the risks could not be alleviated. Telebraun had to resort to a more viscous oil that was not so flammable. Straightaway, though, this oil played a queer trick. In a heavy sea the ship rolled so far that oil slopped out of the containers and onto the floor, thoroughly lubricating the linoleum. Friction was reduced to zero. The heavy battery boxes stored under the seats tore loose and began sliding from side to side as the ship moved, smashing the wall to bits. Zenneck and a sailor tried to tie the skidding boxes down, but their feet slid out from under them, and it was all they could do to save themselves. The salon was a shambles.[135]

Developmental work continued in Strasbourg under less trying conditions. After improving the transmitter, Braun turned to the receiver, which contained the highly unreliable coherer. "Braun suffered so much from its capriciousness," Zenneck reported, "that he was wont to apply epithets to it that consideration for those who are not from southern Germany and thus are unused to strong language leads me to suppress."[131]

Everyone concerned with radiotelegraphy tried to improve the coherer. Its construction had developed into a sort of secret art— a circumstance that came close to wrecking Zenneck's examination for the teacher's certificate. During the physics test, as he was expounding upon Hertzian waves and wireless telegraphy, "the two examiners, Professors Oberbeck of Tübingen and Koch of the Technical University in Stuttgart, interrupted my recitation, examined the coherer I had brought from Strasbourg, and asked questions about its construction and about our experiences in general with coherers. An animated discussion about this field ensued," so that the examination was nearly forgotten.[135]

Braun was already thinking about getting rid of the coherer. Ever since his investigation in Leipzig of the rectifier effect in semiconductors,

there remained with me a feeling of dissatisfaction, and with it, a faint memory that obviously had never died, but remained half-somnolent at the back of my mind. Instinctively I was driven back to this valve effect (with which I had repeatedly, though in vain, attempted to obtain direct current from oscillations of light) when I began to occupy myself with wireless telegraphy in 1898.[96]

The result was that in 1899 Braun did use the rectifier effect to build a crystal detector. But the device offered no advantage over the coherer in the recording of code messages on a moving paper strip, the method then in use. Not until 1901, when audible signals came to be used, did Braun return to the crystal detector and recognize its utility for this purpose. This device may be regarded as a collateral ancestor of the transistor.

Another problem of wireless telegraphy that occupied Braun in 1899 was replacement of the quickly decaying (that is, highly "damped") spark-gap oscillations by more continuous "undamped" oscillations. No way of dispensing with the spark gap had been discovered; yet Braun anticipated that sooner or later methods for producing undamped oscillations would be discovered and that the spark gap ultimately would come to the end of its usefulness for the transmission of information.[135] He therefore found the expression *Funkentelegraphie* ("spark telegraphy"), first employed by Slaby in 1897,[353] quite inappropriate; it is not surprising that when Braun officially became part of Funkentelegraphie GmbH Köln, the word "spark" was soon dropped from the name. (On a later occasion he said, "To call it 'spark telegraphy' describes the principles of operation about as well as calling the stage lighting in a theater 'a heating system.'"[86] But his protests were in vain and "spark telegraphy" it became, at least in German, where it survives in such terms as *Rundfunk* for broadcasting.)

As early as 1899 Braun had been trying to do away with the spark. At a physics colloquium in Strasbourg he proposed to produce undamped electrical waves by means of an ac generator.[135] (The idea was ultimately realized by R. A. Fessenden and E. W. F. Alexanderson in America and by Rudolf Goldschmidt, then a *Privatdozent* in Darmstadt in Germany.) Braun even toyed with the idea of an arrangement that would make use of his cathode-ray oscilloscope; he had a special tube built for the purpose and made some preliminary experiments with it, without success. (A Viennese engineer, Alexander Meissner, ultimately introduced the electron-tube oscillator with feedback as a continuous-wave generator in 1910.)

But Braun had no time to work on his own ideas. Telebraun first wanted to perfect the existing "Braun transmitter" with spark-gap generation. This project occupied Braun throughout the winter of 1899–1900. Not until March 1900 did he get a chance for some relaxation. With a Strasbourg colleague, the botanist Count Hermann zu Solms-Laubach, he traveled through Algeria into the Sahara painting watercolors of mountains, palm groves, and mosques in Biskra and El Guelma.[303]

Braun never said anything outside the walls of the Strasbourg Institute about his ideas and experiments for the generation of undamped waves. "He was not one," as Zenneck wrote later, "to publish ideas or to patent pious hopes."[131] He abandoned this principle only once. In the hectic months of company reorganization in 1898 he sent one idea to the patent attorney Pieper in Berlin "even though it had not yet withstood a real test, because I wanted to be prepared for all eventualities."[307] This idea was the so-called energy switching circuit by which Braun hoped to increase transmitter power by connecting several circuits to a single antenna. Even while the proceedings were still in progress, experiments showed that the patent ought not to be pursued. It was issued anyway on 26 January 1899 (German Patent No. 109,378, "A Circuit to Strengthen Electrical Waves"), and it continued to haunt its inventor. (He did ultimately succeed, in 1905, in discharging the several circuits at the requisite intervals of 1/100,000,000 seconds; but "Wien's quenched-spark transmitter," which came into use in 1908 and was, according to Arco, a useful application of the same principle, soon made the patent obsolete.

The principal improvement made in 1899 involved tuning the receiver to the transmitter's wavelength. But this improvement was first made by others, including Lodge, Marconi, John Stone, and Slaby.[72] Even though he had successfully utilized resonance phenomena in his own circuits, Braun did not concentrate on this aspect of the apparatus. He thought the highest priority should go to the generation of undamped waves—and then one could turn to receiver tuning. Marconi, unburdened by theory, "had recognized its utility before the idea had occurred to me," as Braun put it—one of the many instances in radiotelegraphy in which the method of trial and error forced on an inventor by his lack of scientific knowledge proved to be more successful than the scientific approach.[71]

In December 1899, in the midst of the promising trials aboard the *Silvana*, the Telebraun management, without consultation, ordered that the lightship *Elbe II*, F. L. A. Nietmann captain, be equipped with wireless communication to the land. Zenneck and his carpenter Klaus Heinsohn were requested to make the transfer. The company wanted to attract the attention of business by this first regular use of wireless. It urged that it would make a very welcome Christmas present for the Hanseatic city of Hamburg, which had repeatedly attempted to establish a cable connection between its lightships and the shore, only to have these cables torn apart within hours by the strong tides at the mouth of the Elbe.

It was an expensive Christmas present. Everybody in the little workshop in Döse had to lend a hand to complete the necessary apparatus. This work indefinitely postponed the time when the island of Helgoland could be reached by radio. As this might have been accomplished by the end of winter, Zenneck protested against the directive from Hamburg, to no effect.[135] Equip the lightship first, came the reply; then you can resume work on the Helgoland connection.

While waiting for the extra sets to be fabricated, Zenneck experimented with directional transmission. At the shoreline lighthouse, he had three transmission towers erected in a row and found that transmission was much enhanced in line with the three towers in one direction or the other, depending on which tower was grounded. With the increased signal strength, it was found that houses, trees, or small hills did not interfere with reception. Meanwhile the work of equipping the lightship went slowly forward. Christmas came and went. So did the funds. By Shrove Tuesday of 1900 the company's coffers were as good as empty.

Once again it was Ludwig Stollwerck who "stoked the fires."[307] With a few friends from Telebraun and from Elektrizitäts-AG (formerly Schuckert & Co.) in Nürnberg he formed a holding company with a capitalization of 180,000 marks, of which Schuckert provided 50,000 marks. The practical result was that Schuckert took over the fabrication of the apparatus for the lightships and the Telebraun team in Cuxhaven could return to its experiments with the *Silvana*. But the connection with the high-power engineers from Nürnberg, desirable as it was from a financial viewpoint, had some unexpected consequences. One of the transmitters sent from Nürnberg, in which the Schuckert people replaced the unwieldy Leyden jars by more modern condensers, simply would not work. The Schuckert installation was designed for the low frequencies of ac power transmission and behaved quite differently in high-frequency operation. Another transmitter built in Nürnberg proved to be worthless because the Schuckert designers had thought to save space by placing the condensers inside the coils, without paying any attention to the high-frequency behavior of such circuits—"an arrangement," as Zenneck wrote later, "that was better suited for space heating by eddy currents than for wireless telegraphy."[135]

In an effort to reduce the time required for the work on the lightships, Zenneck had acquired a sailboat in Hamburg at the beginning of July. His aim was to cut down travel time among the several stations, some of which were no more than a few kilometers apart but could be reached only by carts, horse-drawn wagons, and pilot boats in journeys that sometimes took several days. As a member of the Naval Reserve, Zenneck was an experienced sailor, but nevertheless he had let himself in for a dangerous adventure. The first mishap occurred when the boat capsized near Glücksburg and the shipwrecked sailors, Zenneck and his assistant, Dr. Hudtwalker (who could not swim), were run down by the steamer *Patria,* which was coming up the estuary. The steamer immediately lowered a lifeboat and hauled them to safety,[277] but each time after that when Zenneck got into his boat, the sailors in Cuxhaven feared he was "out to drown himself."[135]

In that year Friedrich Niess, who had been one of the founders of the enterprise, died. Niess had played a major role in drawing Braun's attention to telegraphy.

In the same year, Braun was elected a dean in the university. He presided over the faculty of mathematics and natural science from the summer of 1900 until the summer of 1901. The faculty consisted of ten full professors, twelve associate professors, and five *Privatdozenten* in seven institutes (mathematics, physics, chemistry, geology, botany, zoology, and astronomy). The physics institute was assigned a new associate professorship that year to which Mathias Cantor was appointed.

By September 1900 the long-sought connection with Helgoland had been achieved at last. A wooden mast 31 meters high stood on the side of the island facing the mainland; a wooden shack was nearby.[277] On 24 September 1900, Braun was in the shack with Bargmann and the naval commandant of Helgoland, Admiral von Schuckmann, when the following message was received: SEAMAN BECKER HAS RECEIVED SUMMONS BUT CANNOT RETURN TO LAND SO QUICKLY SIGNED NIETMANN CAPTAIN ELBE III.[68] The message was intended for the district court at Cuxhaven. Braun used the incident to demonstrate to the admiral that unciphered messages could not be kept secret.[86]

Braun was on Helgoland because arrangements had been made to demonstrate the connection with the mainland by sending the first formal telegram. At the mainland station, Zenneck and two high officials from the post office department manned the receiving station.[277] Helgoland reception of signals from the mainland and from the lightship had been repeatedly confirmed and checked during the preceding days. Now came the final test: a continuous, distortion-free connection between the island and the mainland.

The test was a great success. After an exchange of preliminary messages, the mainland station asked Helgoland to send a telegram of fifteen to twenty-five words. Characteristically, Braun sent no "congratulatory message meant for the annals of wireless telegraphy" (Zenneck), but rather an improvised poem meant only for the small group of jubilant workers and their guests at both ends of the connection, who were being brought together for the first time through the new electromagnetic medium:

Zum heutigen Feste	For today's celebrations
der Wünsche beste.	Best congratulations.
Trinkt nicht zu viel bei Dölle	Make sure your thirst you've
sonst werdet ihr völle.	mastered
	Lest you get plastered.

The message was received without a single error. A year later Braun spoke about this achievement to a Hamburg convention of natural scientists from all over the world. "As a most severe test, a number of letters that made no sense were transmitted along with some simple and complicated signals. They were transmitted back without a mistake. After this, long and highly modern poems were sent."[69]

With these experiments Telebraun completed its experimental phase and entered on its career as a commercial undertaking. The experimental apparatus at Cuxhaven was transformed into a permanent installation that established continuous contact with the lightboats and with several pilot boats. Within a month, on 29 October 1900, the installation showed how quickly it could respond in an emergency. The four-masted sailboat *H. Bischoff* ran aground during a night storm and the lightship *Elbe II* summoned help by radio.[277] (Unfortunately, the rescue did not succeed: the boat dispatched by the lightship capsized and went down with all hands, and twenty-one of the twenty-nine-man crew of the sailboat also drowned.[125])

The news that Telebraun had equipped pilot boats with radiotelegraphy and had established contact with Helgoland brought the company offers and inquiries from all over the world. Could "Prof. Braun's system" be installed in Scotland, in Scheveningen, on the Adriatic Coast?[277] Finally there were signs of a large amount of business, but Telebraun had no money to compete for it. All its cash, including Schuckert's recent contribution, had been literally "thrown into the air," and it would be difficult to put together even another 100,000 marks. Big business was not yet interested. Postal services and governments were hesitant because of the political and financial influence of the international cable companies and their own large investments in cable telegraphy. A single transatlantic cable might cost as much as 45 million marks, yet the amount had been found; but could the same backing be expected for radiotelegraphy?[235]

Telebraun tried to interest the powerful Hamburg shipowner Albert Ballin, of the Hapag line. Ballin might have risked it,[307] but declined when he learned about the connection with Schuckert. What would Berlin say? In Berlin, the official expert was Slaby, the Kaiser's advisor. Ballin reasoned (doubtless correctly) that nothing could come of a radiotelegraphy project without government support—surely the military would be important customers? If he were to go into radiotelegraphy, said Ballin, it would be more prudent to stick with Slaby. Braun's "coupled transmitter" was all well and good—but what really mattered at the moment were influential connections. Both Zobel and Stollwerck's personal emissary Emil Heimerdinger came back emptyhanded from interviews with Ballin, who proceeded to equip the *Deutschland,* the flagship of his fleet, with radiotelegraphic apparatus by Slaby.[307]

Another blow was the news that Emil Rathenau, the director of AEG and a confidant of the Kaiser, had entered on an agreement for the mutual use of his patents and experience. Not just little Telebraun but even the mighty Marconi Wireless Telegraph Co. Ltd., which had been recently formed in London, could hope to prevail against the new constellation—Slaby's position at court, the industrial might of AEG, and the well-equipped AEG radiotelegraphic laboratory directed by Slaby's former assistant Georg von Arco. If Telebraun wanted to survive, it would have to find a new solution, and quickly. The obvious thing would be to turn the flirtation with Schuckert into a marriage—but with an even stronger partner, Siemens in Berlin.

Siemens seemed predestined to play the role of bridegroom, if only because it had already undertaken radiotelegraphy experiments of its own. Moreover, its financial strength made it a natural competitor of the AEG. The merger between Telebraun and Siemens was a logical new alliance.

At this point Braun was given his first opportunity to break his silence. He never spoke about what it had meant to him, as a working scientist accustomed to making new-found knowledge promptly and generally available, to remain silent for twenty-three months, a period in which each day brought new insights to be shared with those who were floundering about in the new field. He had been able to demonstrate his attitude toward pure science when he published detailed descriptions of his cathode-ray oscilloscope. But with the radiotelegraphy venture, patent attorney Pieper saw to it that not a word was printed.[307] Braun was not even permitted to correct the errors that appeared in the vague accounts by reporters who got their information from assistants and the Cuxhaven mechanics.

Now everything changed. To establish the superiority of his system, Braun was urged to speak out. At this juncture silence would be damaging. The business world must have seemed quite topsy-turvy to him.

The manuscript for the first public position paper had already been drafted.[68] The new stage, in which the watchword was maximum publicity in the form of lectures, scientific publications, and newspaper articles, began on 16 November 1900, in Strasbourg, the home of "Braun telegraphy." It was a public lecture sponsored by the Strasbourg Science Union and delivered before a large audience in the lecture hall of the Strasbourg Physics Institute.[282] Both foreign and domestic newspapers reported the event in long articles.

Braun began with a historical introduction about wireless telegraphy in general. Detailed explanations of water telegraphy followed. He noted that Cantor had repeated the early experiments in the moats of Strasbourg and again in the salt water of the North Sea, where distances up to 3 kilometers had been spanned. Braun then turned to "wireless telegraphy through air." He explained his transmitter, showed illustrations of his patents, and conducted several experiments: he set off long and short sparks; he silenced the high-voltage antenna of Marconi's original transmitter by touching it with a wet, grounded string; he demonstrated the effect of resonant tuning by lighting Geissler tubes; and he showed that his transmitter was perfectly safe by drawing long bright sparks from the antenna that could scarcely be felt.[282]

Braun concluded that radiotelegraphy would not eliminate cable telegraphy, but that it had a clear place in signal transmission, military applications, and "thinly populated areas where a telegraph line is exposed to dangers from storms, wild animals, or ignorant people." He ended modestly: "We offer our best wishes to the child in its cradle. We are delighted when he seems to be developing in accordance with our wishes. Yet who can say with certainty after only five years what he will be like when he becomes a man? He will grow, and perhaps he will achieve something even if he does not turn out to be a Hercules."[68]

In December 1900 the merger was completed. With the formation of the AEG-Slaby-Arco group, the Siemens people also became anxious not to be outdistanced.[235] But it must have taken a skillful effort to convince Siemens to join a group working in electricity that was headed by a chocolate manufacturer! Schuckert & Co., conceding its lack of expertise in telegraphy,[307] broke off its participation and was replaced by the experienced telegraphy branch of Siemens & Halske AG of Berlin, which joined with Telebraun to form a new venture: Gesellschaft für drahtlose Telegraphie, System Prof. Braun und Siemens & Halske mbH, Berlin. Telebraun's contributions were the Braun patents in several countries and the practical experience of two years of experimentation. Siemens & Halske contributed its own division for research and development under the direction of the physicist Dr. Köpsel.

The joint venture undertook to purchase all its equipment from Siemens & Halske. In return, the Siemens exchequer, worldwide business connections, and production capability were now at Braun's disposal. Braun was also able to use some valuable products of the Siemens laboratory, including the most reliable coherer available at the time; the "variable condenser," one of the most important components in high-frequency engineering; and an excellent telephone receiver—all constructed by Dr. Köpsel—to say nothing of excellent laboratory facilities in various locations, both near the sea and in the mountains.[276] Siemens branches all over the world started handling the inquiries and requests for quotations in a way that Telebraun could never have managed by itself.

The price of all these new advantages was a substantial loss of the original participants' control. For all practical purposes, the Braun-Siemens company became a subsidiary of Siemens & Halske.[235] The original participants were not sorry to give up the unequal struggle and were glad to leave the management of the new company to the seasoned professionals of Siemens. Bargmann remained as nominal manager and Stollwerck, Braun, Zobel, and the Hamburg attorney A. H. Kleinschmidt were board members.[307] But everyone understood that the actual direction was firmly in the hands of board chairman Wilhelm von Siemens and the managing directors of the Berlin plant, Drs. Raps and Zimmer.

Telebraun itself retained an independent existence until 1908. Its only annual task was to distribute the Braun-Siemens dividends to the Telebraun shareholders. There were seven major shareholders with shares worth 100,000 marks or more (Stollwerck, Braun, Zobel, Schmitz-Schlagloth, the Niess estate, Gümbel, and Wehner) and sixteen smaller shareholders, among whom were Prof. A. Voller of Hamburg, the utilities executive Franz Joly of Cologne, Carl Pieper of Berlin, as well as Prof. Mathias Cantor and Dr. Jonathan Zenneck, who had received shares as rewards for their participation in the Cuxhaven experiments.[307]

The yield of the Telebraun shares varied. The dividends for 1902 were 2.5 percent; for 1905, 24 percent. There were also years without any distribution whatsoever. One Telebraun pioneer did not long enjoy the success for which he had fought so hard. Albert Zobel, who had gone from merchant to banker to chairman of the board of Telebraun, died in Giessen in 1911 at the early age of 45.

The early days of radiotelegraphy were marked by mutual recriminations and patent interference suits as various experimenters worked their way toward similar solutions. Oliver Lodge and Alexander Muirhead (who had formed a syndicate) accused Marconi of having appropriated their ideas on tuning. They filed suit. The case dragged on for years and ultimately was settled out of court in an agreement by which the Marconi company acquired the Lodge-Muirhead patents.[356]

In a similar way, Braun felt that Marconi simply had helped himself to Braun's coupling circuit, even though it was protected by British Patent No. 1862 of 26 January 1899.[72] Braun said that Marconi, when the matter had been broached, had admitted the borrowing "with commendable frankness." Braun felt that Marconi's patent of 26 April 1900 (No. 7777, the famous "four-sevens" patent) bore a strong resemblance to the first part of Braun's British patent, and that Marconi's patents of 7 January 1901 (Nos. 409, 410, and 411) resembled the second part of Braun's patent. Only the third part of his patent, Braun noted caustically three years later, was still "free" in Britain. But he did not sue.

Marconi's partisans saw the matter differently. "It has sometimes been suggested," wrote Marconi's long-time collaborator J. Ambrose Fleming in 1906, "that Marconi availed himself of Braun's prior invention, but in truth his (Marconi's) investigations were carried out quite independently, and conducted to a more practical issue than those of Braun—at least up to the date when the latter secured his first German and equivalent British patent, No. 1862, of January 26, 1899."[357]

As a scientist, Braun could perhaps take pleasure in the successes achieved by Marconi with circuits Braun regarded as his own. (Marconi's new distance record of 297 kilometers in January 1901, soon after Braun's Helgoland connection of 63 kilometers, received worldwide publicity.) For the directors of Braun-Siemens, however, such thoughts held little consolation. Yet they took their time about bringing suit against Marconi, and this delay later proved very costly.

The directors were actually more concerned over their home-grown competitor, Slaby, in whose circuits Braun's arrangements began to appear more and more frequently. The problem was recognized quickly, but the twenty-three-month delay between the conception of Braun's circuit and its announcement now proved to be a great disadvantage. Many scientists simply had no idea that some of the discoveries being claimed by Slaby had been described by Braun's patents of 1898. His colleagues, said Braun, "were not familiar with my methods, which can be explained quite simply by the fact that I have publicized them only very occasionally." He also noted that in Slaby's Berlin lectures "he never even mentioned my name in those places where the scientific world customarily might have expected him to do so."[77] Slaby simply pretended that Braun did not exist.

Braun responded by publishing an article in the *Physikalische Zeitschrift* on 5 March 1901, "On Rational Transmitter Circuits in Wireless Telegraphy," a review paper intended to serve as an introduction for the complete neophyte. A more detailed description and explanation of his own patent followed on 22 March 1901. To make sure that it would not be overlooked this time, he published it in the *Elektrotechnische Zeitschrift*, the journal of an engineering society under the influence of Slaby.

More difficult to counter than Slaby's scientific pretensions was the increasing resemblance of his transmitter to Braun's. Slaby proposed transmitter circuits that at first glance seemed to have nothing in common with Braun's. But just as the parts of a prefabricated house must always form the same house, so even Slaby's most complicated arrangements could be shown to be identical with Braun's patented circuits.

In fairness to Slaby one must consider the possibility that he did not recognize these identities and that he thought of each modification, no matter how slight, as representing a new circuit. Referring to a lecture Slaby gave on 10 June 1901, Braun remarked that "Herr Slaby is not clear about the function and the value of the condenser circuit."[74] That was Slaby's personal tragedy: doubtless, he had earned a place in the annals of radio "by virtue of his great diligence and aptitude in seizing opportunities as they presented themselves;"[184] yet in the judgment of an early account of the pioneers of radio published in Germany in 1926, he "never made any truly fundamental discoveries."[273]

In retrospect, Braun emerges from the dispute as the one scientist who actually understood the principles of radiotelegraphy; the others seemed to be groping in a fog of ignorance, error, and arrogance. But it was precisely his knowledge that made him recognize the ever present danger of a deliberate modification or accidental "reinvention" of his circuit. He wrote later that in his paper, "On Some Alternative Connections for Wireless Telegraphy,"[71] he had described several possible variations and combinations largely in anticipation of other possible circuits that might turn out to be difficult to attack in the patent courts.

Slaby's development of "spark telegraphy" sailed ahead in close cooperation with the navy. The scales of Justice were partially balanced by the good relations that the Siemens company enjoyed with the army,[235] whose first mobile radiotelegraphy installations had come from Siemens in 1901. They were installed on horse-drawn carriages and the antennas were carried aloft by balloons; in windy weather, kites were used. The military gazette *Militärwochenblatt* reported that the Braun-Siemens system had performed "extraordinarily well" at the 1901 imperial maneuvers. Messages were transmitted reliably over more than 100 kilometers.

Radiation of the transmissions in all directions was a problem in military applications, since it allowed the enemy easy access to them. In the summer of 1901 Ferdinand Braun received an army contract to develop a "directional wireless telegraphy." With a Prussian officer, Captain von Sigsfeld, he conducted experiments on the Strasbourg drill ground in which Zenneck's Cuxhaven arrangement was perfected to such an extent that radiation in certain directions ceased almost completely (see figure XIII in appendix B). But a patent application revealed that an Englishman, S. G. Brown,[358] had had this procedure protected by a patent two years earlier.[95]

In the course of these experiments, von Sigsfeld had observed that "certain atmospheric disturbances noticed in the receivers repeatedly coincided with variations in the earth's magnetic field; he was inclined to view them as electrical waves originating in space, notably from the sun."[78] Von Sigsfeld died the following year without having had an opportunity to test what Braun later called "his bold hypothesis." This hypothesis in fact later was confirmed through observations of the effect of the sun on events in the earth's atmosphere.

Radio science was in a sense born in 1901 at the Physics Institute in Strasbourg (figure 25). Until then, most physicists tended to

Figure 25
A new science is born. Jonathan Zenneck, Ferdinand Braun, and
Mathias Cantor (right) experimenting with high-frequency oscillators at
the Strasbourg laboratory.

regard it as part of electrical engineering. The narrowness of this
view was first demonstrated in Braun's lecture before the Association of German Scientists and Physicians in Hamburg in
September 1901, at which the well-known Prague expert on electromagnetic oscillations, Ernst Lecher, joined in the discussion.[69]
Braun made his point even more strongly with a theoretical piece
published in the *Annalen der Physik* on 20 March 1902, "On the
Excitation of Standing Electrical Waves along Wires by Condenser
Discharge." The problems arising from radiotelegraphy became
the foundation for a new science, high-frequency physics (the
study of rapid electrical oscillations), which soon developed into
a legitimate branch of physics through the efforts of many capable
scientists, notably Max Wien in Danzig. High-frequency physics
was to play a part in many branches of science and to engender
several branches of technology.

Braun was strongly supported in his efforts to raise the status of radiotelegraphy to that of a scientifically respectable subject by the veterans of the Cuxhaven waterfront experiments, Mathias Cantor and Jonathan Zenneck. Zenneck's 1019-page book *Electromagnetic Oscillations and Wireless Telegraphy* was the first systematic treatment of the new branch of science. His *Habilitation* dissertation (to qualify for faculty status) was on a related subject and enabled him to make Strasbourg the first German university to offer a course of lectures on electromagnetic oscillations, in the winter of 1901–1902.

Zenneck intended to show some photographs in the course of the formal *Habilitation* lecture. He placed them in the pocket of his tailcoat. When he needed them, though, he just could not find the pocket. Without regard for the solemn assembly or for the dignity of his own office, Braun sprang to his aid, pulled up Zenneck's tails, found the pocket, and produced the photographs amid general hilarity.[135]

In carrying out his mainfold radiotelegraphy projects, Braun had several private assistants in Strasbourg. The master was always surrounded by a flock of young people. Most of them may have come to Strasbourg with quite different plans. Once there, however, they became converts to the new science with youthful enthusiasm.

Several alumni of this group later became important high-frequency physicists in their own right, notably two Russians, Mandelstam and Papalexi. Leonid Isakovich Mandelstam, born near Odessa, had come to Strasbourg in 1899 at the age of 20 after being expelled from the University of Odessa for his participation in revolutionary student disturbances.[135] Braun directed the young Russian's energies toward a new field that soon turned out to be of the utmost importance: the development of methods of measurement for radiotelegraphy. With his 1902 dissertation, "Determination of the Period of Oscillation of an Oscillatory Condenser Discharge," Mandelstam became the first doctor of the new science.

The problem of monitoring high-frequency oscillations had been dealt with until then by means of the rotating mirror introduced by Feddersen in the course of his pioneering condenser-discharge experiments some forty years earlier. Mandelstam tried to use Braun's cathode-ray oscilloscope, which had performed quite satisfactorily at the much lower ac power frequencies, but no one then knew how to make the instrument work at the higher frequencies. In the end he used an ingenious method proposed by Braun, based on the entirely different calorimetric principle: measurement of the heating of an induction coil under the action of alternating current, comparison with equivalent heating obtained in an ohmic resistor, and computation of the "inductive resistance" of the coil from the comparative measurements. For a given condenser size, the frequency and hence the wavelength could then be readily determined.[285] Braun was annoyed by a defect in this method, since a spark had to be generated for each measurement, but he admitted that it provided a convenient way of measuring the frequency in a matter of minutes, a significant advance.

The other young Russian, Nikolai Dimitrievich Papalexi, had matriculated at Strasbourg in 1900. He was the son of a Crimean landowner, a year younger than Mandelstam, who had spent a semester in Berlin before coming to Strasbourg. Papalexi became great friends with Mandelstam; the two young men, according to Zenneck "both exceedingly bright and diligent high-frequency physicists," became inseparable both as friends and as scientists. When the German university at Strasbourg was closed at the end of World War I, Papalexi first strove to continue the Braun school elsewhere in Germany. But, as a Russian exile, he found all doors closed to him. Meanwhile, Mandelstam had returned to the Soviet Union and had been launched on a brilliant career there. It was through his intervention that the "bourgeois" Papalexi received permission to return to Russia. The two friends continued the "Braun school" there. (As a unique memorial to Braun, it is interesting to find both listed among the "fathers" of the first man-made satellite in *Pravda* of 16 October 1957, two weeks after the launch of Sputnik I.)

In 1928 Mandelstam and Papalexi published a tribute to their teacher in *Die Naturwissenschaften* on the tenth anniversary of his death. They spoke of his "path-finding activity in the field of electrical oscillations" and characterized it as "basic to the modern development of radio technology." They went on to say,

Ferdinand Braun worked creatively and keenly in the entire field of radiotelegraphy, and his name will always be among the first in the history of the development of this splendid field of applied physics. . . . As a teacher, he will remain unforgettable to all who had the good fortune to work in his institute. He had the ability to anticipate phenomena, and he possessed a remarkable feeling for the art of experimentation—things that cannot be taught and that are granted to only a few select researchers. These qualities, to which in large part he owed his zest for research, also benefited his students.

Braun's research accomplishments have assured him a high place of honor in science. All who had the good fortune to be near him are sure to remember this great, wise, experienced, and benevolent man with love and respect.[123]

The fourth year of radiotelegraphy ended with a major triumph: on 12 December 1901, Marconi succeeded in bridging the Atlantic by radio for the first time. He had erected a transmitter, constructed on a larger scale than anything previously attempted,[283] in Poldhu in Cornwall, near the westernmost point of England; and had received signals with an improvised receiver near St. John's on the east coast of Newfoundland. Both transmitter and receiver resembled Braun's circuits. The prearranged transmissions were, said Marconi, clearly received.

This announcement created a world sensation.[359] The news that a distance of 3,470 kilometers between the two continents had been bridged overshadowed all other news about radiotelegraphy, including reports on the intensifying commercial competition in Germany. Braun-Siemens in Berlin issued a statement that the long overdue patent infringement suit against Marconi would now be initiated and that Prof. Braun "on his part was preparing to bridge the ocean with his long waves."[307] Marconi's success had made a sufficiently strong impression on the two rival German groups to cause them to come together in Kiel to discuss a merger between AEG-Slaby-Arco and Braun-Siemens. But nothing came of any of it. The suit against Marconi did not get under way; Braun did not bridge the ocean; and the Kiel negotiations, begun with much hope at the beginning of February 1902, were suspended in March. Braun would have been willing to come to an agreement with his rivals, but Slaby's insistence that the new concern be continued under his name alone was unacceptable to Braun. The only winners were the Marconi interests, which now could establish an effective world monopoly.

In the year that followed the failure of the merger talks at Kiel, the hostility of the two German systems became a matter of national concern. The judiciary, politicians, and the public were all occupied with it. The names of Braun and Slaby became well known to the man on the street, for the struggle was portrayed as one between the two scientists rather than between the two giant companies, which continued to maintain a respectful attitude toward each other.

This was the year in which the 52-year-old Braun's hair turned gray and in which Stollwerck came to regard wireless as his "child of pain."[268] The legal situation was favorable to Braun-Siemens.[235] In the autumn of 1901 the AEG-Slaby-Arco group had unsuccessfully attempted to patent a circuit of Slaby's that turned out to be a disguised version of one of Braun's circuits. In May 1902 Slaby attempted an interference suit against Braun's patent, contending that his laboratory notebooks of 1898 showed him to have anticipated Braun's circuit. This maneuver was also unsuccessful. The court found that a claim based on this unpublished material was inadmissible and sustained Braun's argument that "even if Herr Slaby . . . had experimented with the circuit he sketched in the autumn of 1898, he clearly did not recognize the importance of the arrangement. For a long time he failed to follow the path he had accidentally crossed."[76]

The AEG-Slaby-Arco group next tried to establish the priority of a circuit published in the summer of 1898 by the Berlin physicist Martin Tietz in the *Elektrotechnische Zeitschrift*. Again the suit was unsuccessful; the court held that the circuit was not identical with Braun's. Tietz had placed only a condenser across the spark gap and not a complete tuned circuit comprising both condenser and induction coil.[76]

At this point a series of events that had begun in December 1899 produced a major victory for AEG-Slaby-Arco. It started with the sensational wreck of the Russian cruiser *Apraksin* on the open sea in the Baltic. That led the Czar, Nicholas II, to order his navy equipped with the new telecommunications system. (Popov was still conducting his private experiments at the time.) This advance alarmed the Kaiser, who complained that his navy had been experimenting with Slaby's apparatus for the past two years and had still not made up its mind between that system and

Braun's. He ordered a competitive trial of the two systems in the Baltic Sea.

Slaby was well prepared. His circuits had grown increasingly similar to Braun's. Indeed his latest transmitter was a faithful copy of Braun's patent of 14 October 1898. The two years of experimenting on naval vessels had given Slaby an added advantage.[286] Braun-Siemens was given exactly two days to install its equipment on the ships and to conduct some preliminary tests. They achieved a range of 105 kilometers, while Slaby's apparatus achieved 115 kilometers. Braun-Siemens argued in vain that Slaby had infringed its patent. The Kaiser was interested only in range (figure 26).

Braun-Siemens decided that its only alternative was to go to court. The first hint of the outcome of this suit was the newspaper reports emanating from the 1902 meeting of the Association of German Scientists and Physicians at Carlsbad, where Slaby's conduct was openly criticized by several reputable scientists, among them Max Wien, whose just published theoretical paper on ra-

Figure 26
Apparatus used in demonstration for Kaiser Wilhelm II; left, Braun transmitter.

diotelegraphy had attracted a great deal of attention.[284] One commentator went so far as to predict a victory for Braun-Siemens but that the navy nevertheless would be forced to go back to Slaby's "obsolete, no longer competitive system."[286] Far-reaching consequences for military and foreign policy were predicted.

At the Kaiser's court a storm began to brew, but the "supreme displeasure" was directed less at AEG-Slaby-Arco than at Braun-Siemens. The shipowner Ballin's prediction that the Kaiser would stand by Slaby was coming true. This show of favor encouraged Slaby to take a rash step, attacking Braun's claims of priority in the pages of the *Annalen der Physik*. His arguments were essentially the same as the ones that had already been rejected by the courts, though, and they were torn apart by Braun in two articles published in the *Annalen* in October and December of 1902.[74, 76]

AEG-Slaby-Arco now took refuge in the increasing press agitation for a merger between the two groups. A merger would also have pleased the bureaucracy, which would have preferred to deal with only one enterprise. Now, however, Braun-Siemens was not interested in a merger. Once the courts decided the infringement proceedings in its favor, the company would dominate radiotelegraphy throughout Germany. "The strongest man in the world is he who stands most alone," wrote Georg Siemens later, quoting Ibsen's "An Enemy of the People."[235] But they were to learn quickly that other powerful forces were at play.

Why had the proceedings against Marconi been postponed again and again? One commentator, Otto Jentsch, suggested in January 1903 that Braun-Siemens, discouraged by the failure of the Kiel merger negotiations, might be considering a merger with the Marconi interests instead.[286] His speculations may not have been wholly unfounded. A letter from Braun in Strasbourg dated 15 January 1901 and addressed to an unnamed Herr Commercienrath (almost certainly Stollwerck) indicates that he was considering such a merger even then as an alternative to an agreement with Siemens. He writes,

M. has an unmistakable start on us from the practical and commercial point of view. If we continue to experiment for some time longer—of course with a corresponding expenditure of funds—then it might turn out that negotiations with M. could come under consideration later. But doubtless the most desirable move would be a merger as soon as possible.[104]

In his comments Jentsch makes an argument that well represents the sort of pressure being brought to bear on Braun-Siemens:

Such a merger [with Marconi's] would be contrary to German interests. For example, the Marconi Co. might obtain a monopoly if the patent proceedings between the Braun-Siemens and Slaby-Arco systems should result in a victory for Braun-Siemens—which, as far as one can tell now, seems quite likely. . . . In the interest of a continuing calm and fruitful development, let the two parties bury the hatchet and proceed jointly against the Marconi Co.—'tis a consummation devoutly to be wish'd.[286]

Although far removed geographically, Braun in Strasbourg was centrally involved in these events. He was now a public figure, much discussed in the newspapers, still hoping for a favorable outcome because he relied on his patents and on constitutional law. His sense of humor remained unimpaired. Radium therapy had just come into use and his son Konrad had received radiation treatments that cured him of a peculiar ailment: he had been unable to bend his spine. "Not even before the high and highest nobility?" Braun asked his son during a party in his home one evening. "No, not even before them," replied Konrad. "Seems to be a congenital defect," said Braun.[325]

During the summer of 1902 Braun continued the experiments on directional transmission that he had started the previous year with Captain von Sigsfeld. The results did not prove very suitable for use in a tactical military situation marked by rapidly changing topographical conditions. Moreover, the airship division of the Royal Prussian Army expressed misgivings because of the height of the antenna. A new method was attempted, involving a horizontal antenna that inclined moderately downward at one end. That worked after a fashion but still fell short of being a solution "on the thorny path to highly directional radiotelegraphy," as Braun put it.

Other experiments were conducted by Mandelstam and another research worker at the institute, Hermann Brandes.[95] They investigated the effects of inductively coupled antennas on the receivers and obtained the rather surprising result that receiver performance improved when the coupling was made looser. A byproduct of this observation was the development of a new type of instrument that is still in use. When Dr. Franke, assistant director of research at Siemens & Halske, looked over the experiments, he suggested that they might be used to develop a helpful engineering device. The project was assigned to an engineer named Dönitz; the resulting device, a wavemeter, was for a time known by his name. That was perhaps the first commercially available meter specifically designed for radiotelegraphy. It consisted simply of a tuned circuit and an indicating needle, the whole loosely coupled to the circuit whose wavelength was to be determined. The wavemeter's components were variable, for instance by means of a rotating-vane condenser; when the needle's maximum deflection indicated resonance, the wavelength could be read off directly. This simple configuration quickly replaced more laborious methods for determining wavelength, such as Mandelstam's calorimetric technique.

Another basic element of high-frequency technology born in Strasbourg in 1902 was the predecessor to present-day ferrite components—coils with cores pressed from a mixture of iron powder and an insulator. Rutherford in England had shown in 1897 that magnetized steel needles were demagnetized by electromagnetic waves. That was also the year Braun had shown how rapid variations in the magnetic excitation of iron could be observed by means of his cathode-ray oscilloscope. The two observations were combined in a doctoral dissertation at Strasbourg in 1901 by an English student, William Mansergh Varley, "On Magnetism Caused in Iron by Rapidly Oscillating Fields."[287] Varley's results caused Braun to assign still another problem that year to Hermann Brandes: to determine whether the magnetic energy liberated when demagnetization occurred as a result of rapid oscillations "might not be usefully applied to reinforcing the excitation of a coherer."[77]

The investigation that followed showed once again that research, once under way, often leads to wholly unanticipated results. Varley's work had been done with wires and bundles of wires. Brandes (like Rutherford) used somewhat thinner steel needles and later the still finer balance spring of a pocketwatch. Braun carried refinement still further, to finely ground iron powder mixed with vaseline oil (an insulator) and compacted into a glass capillary tube—an arrangement suggestive of the later ferrite core. When the tube was inserted into the induction coil of an oscillating circuit, the energy contained in the oscillating circuit rose considerably.[77] All other things being equal, a spark-gap length of 9 millimeters was increased to 12 millimeters by the insertion of the core.

In his description of this series of experiments in the *Annalen der Physik* at the end of 1902, Braun suggested that the ability of cores to concentrate energy might serve to reduce the number of turns in coils, to increase coupling between coils, and to decrease energy losses in coils.[77] Its descendant, the ferrite core, ultimately developed into a most versatile electronic part, widely used in computer memories, antennas, and many other applications.

In April 1903 Kaiser Wilhelm II let it be expressly known that he viewed the rivalry between the two great German concerns, AEG and Siemens & Halske, with great disfavor, especially since it served to strengthen the hand of the Marconi interests in Britain. Under that sort of pressure, a merger was quickly consummated. Siemens & Halske had the stronger legal position but AEG held a trump card: the Kaiser's favor. AEG's director Emil Rathenau had direct access to the Kaiser and might have given him a one-sided report if the merger negotiations had failed.[360] Siemens & Halske saw the handwriting on the wall. At the last minute a way was found to save face: the signatory of the merger agreement, dated 15 May 1903, was not Siemens & Halske, or Braun-Siemens, but the almost forgotten Telebraun.[235]

The four erstwhile independent German radio pioneers—Braun, Slaby, Arco, Siemens—agreed to pool everything: their thirty patents, their thirty-three employees, any business connections, and all their experience. The name of the new company was Gesellschaft für drahtlose Telegraphie mbH. The first two syllables of Telebraun and of Slaby's Funkentelegraphie (spark telegraphy) were used to form the telegraph address of the new firm, Telefunken, which later became the trademark and finally the name of the company.

Within a short time the thirty-three employees of Telefunken grew into an enterprise that could compete worldwide against the Marconi Wireless Telelgraph Company. This was a hard-fought battle that stretched over eight years. Marconi had used his head start well. While the German systems were feuding with each other, Marconi had raised large amounts of cash through a number of stock companies, built his own factories that produced his equipment in large quantities, and, most important, trained radio men who, as "Marconi operators," were rented out with his apparatus. They had instructions not to communicate with any stations of another system.

In view of the dominant position of the British fleet and its network of supply stations, the Marconi world monopoly was nearly unassailable, especially in marine communications. If one views international competition as conducive to the advancement of technology, one must regard the founding of Telefunken as a blessing. Yet for Braun-Siemens the merger meant a considerable initial sacrifice. In the newspaper reports, the basic Braun patent was placed on a par with Slaby's "technical accomplishments in naval radiotelegraphy."[286] The direction of the business was in the hands of Georg Count Arco (figure 27), who came from AEG and who proved to have good business acumen. The names of Slaby and Arco took over the title page of the new firm's an-

Figure 27
Georg Count Arco (1869–1940).

nouncements; the names of Braun and Köpsel did not appear until page 2. As Arco testifies, though, Braun "quickly fitted into the new situation. He placed the prospects for the favorable development of wireless telegraphy through a strengthened new organization above personal interests and wishes and after a short time became an enthusiastic worker in the newly founded company."[114] The two men soon became warm friends, and Braun later arranged for the University of Strasbourg to award Arco an honorary doctorate—the only time throughout his university career that Braun had ever initiated such a proceeding.

As for Slaby, the victory of the merger was not enough. Although now a member of the same organization as Braun, he did not hesitate to mount further attacks against his associate. These sallies ultimately cost Slaby the Kaiser's favor. He died in 1913, a bitter and lonely man.

By 1903 Braun had become a public figure. He was bombarded by requests and invitations. Would he speak at the International Radiotelegraphic Conference in Berlin? Would he give a lecture at Karlsruhe? Was it true that energy could be transmitted in arbitrary quantities over arbitrary distances by Braun's system?

To the last inquiry, which came from Nuremberg, he replied, "I shall believe in that possibility only after it has been demonstrated that a room in Munich can be heated by means of a searchlight located in Nuremberg." Long-distance transmission of electromagnetic energy in quantities sufficient to perform useful work at the receiver had been proposed earlier by Tesla,[361] but Braun saw clearly that it could not be done with the means then available. (Three quarters of a century later, this objective has still not been accomplished, although proposals involving lasers or microwave transmission of solar energy from outer space abound.)

Of more interest was the forthcoming International Radiotelegraphic Conference, at which the British Post Office was expected to press for agreements that would consolidate the British monopoly. Braun listed his views in a newspaper article:

Ships and stations equipped with radiotelegraphy apparatus must be obliged to handle one another's traffic regardless of the system employed.

The wavelengths to be used must be prescribed.

The construction of powerful stations for traffic over long distances that can be bridged by other means should be licensed only if such licensing does not impede more urgent tasks that it is in the nature of radiotelegraphy to perform.

States must eschew regulations that would interfere with such radiotelegraphy experiments as are necessary for the development of the new technology.[75]

This article correctly anticipated the increasing intervention of governments in the development of radiotelegraphy and the resulting limitations on the freedom of researchers to carry out their work.

The Karlsruhe lecture took Braun back to the Technical University whose Physics Institute he had directed a score of years earlier. The new occupant of Braun's old chair, Prof. Lehmann, liked to show experiments on a large scale. In his lecture room, in place of the usual laboratory bench, he had had a stage built on which he mounted his demonstrations.[135] Zenneck later recalled a lecture demonstration made during Braun's visit, which showed that a movable iron core is drawn into a coil when an electrical current passes through it:

A huge cylindrical coil was supported on two blocks with its axis in the vertical direction. A large iron cylinder was placed beneath the coil so that it penetrated partially into it. Affixed to the cylinder and passing through the coil was a large wooden pole, with a seat on top. An assistant was placed in the seat. When the current was turned on, the iron cylinder soared aloft, carrying the assistant into the air.[135]

The available power supply limited the weight of the assistant to 75 kilograms. The story went around among German physicists that when Lehmann needed a new assistant, he would place an advertisement in the *Physikalische Zeitschrift*: "Assistant wanted. Not over 75 kg. Karlsruhe Physics Institute."

The demonstrations in Strasbourg were on a more conventional scale; but the number of students had grown. As a famed research center for the new high-frequency physics, Braun's institute attracted young physicists from all over the world. Between 1901 and 1903 the number of students in the natural sciences increased by 25 percent.[203] At the same time there was a regrettable loss in the faculty. Dr. Mathias Cantor, for thirteen years Braun's faithful follower, was appointed associate professor of theoretical physics at the University of Würzburg in the autumn of 1903. He remained there until the beginning of World War I. In 1916, at the age of 45, he died as a result of combat injuries.

Further progress in high-frequency physics came with two doctoral dissertations. Papalexi wrote one on instrumentation, "A Dynamometer for Rapid Electric Oscillations." Another dissertation, "Damping of Condensers in Spark-Gap Circuits," was written by 22-year-old Georges Rempp, an Alsatian who later became the chief meteorologist of Alsace-Lorraine and until 1937 was professor of meteorology at the French University of Strasbourg. Rempp's dissertation was a solution of a prize problem—a sign that high-frequency physics was carried on in Strasbourg "across the board." Another sign of that was the rejuvenation that took place when Cantor left: Zenneck became first assistant, Brandes second assistant. The third assistant, Mandelstam, was also a high-frequency physicist. Mandelstam replaced Feustel, who had left in the summer of 1903 after completing his dissertation, "On Capillary Constants and Their Determination by the Smallest-Bubble Method."

High-frequency physics never occupied Braun exclusively. He had the happy gift of a great catholicity of scientific interests, which saved him from becoming absorbed in a narrow scientific specialty.[123] Brought up in the natural science tradition of the eighteenth and nineteenth centuries, he would not accord physics, let alone one of its branches, dominion over that whole that he considered natural science to be. He remained true to the principle of the primacy of the universal over the particular. He sought always for connections between physics and other branches of science. Within physics he demonstrated "a rare capacity, found only in unusually gifted persons, for uncovering unifying relationships linking . . . fields that at first seemed to have nothing in common," as Arco expressed it.[114]

As soon as the future of radiotelegraphy seemed assured, Braun started searching for new problems to work on. Even in the most difficult years he had always managed to devote some of his time to totally unrelated problems. There was, for example, a mathematical investigation of the possible effects of gravity on the growth of plant cells "in the form of pressure differentials between the upper and lower side of a cell."[67] He was also closely involved with the experiments on capillarity being done by his assistant Robert Feustel. Braun had been interested in this subject since 1891 when, at Winkelmann's request, he had undertaken the chapter on capillarity in addition to the one on thermoelectricity for the *Handbuch der Physik*. He had discovered then that the published figures for the capillary constants of the most frequently investigated liquids, water and mercury, varied by as much as 25 percent.[40] Feustel succeeded in narrowing this range in an investigation based on a novel method proposed by Braun.

Once things had settled down in radiotelegraphy, Braun's research turned to a general investigation of oscillatory phenomena, a subject in which he had shown an interest as far back as his 1872 doctoral dissertation. Then he had sought connections between optics and acoustics. Now he started to explore theoretical and experimental analogies between acoustics, optics, and electromagnetic waves. This work came to fruition in a number of papers during the years 1903–1905.

A first step was an application of experimental knowledge gained in the work on directional antennas to acoustics. Braun constructed a resonant tube on the basis of his electromagnetic experiments that would enable ships to pinpoint foghorns during foul weather.[78] An attempt to construct a practical instrument fell victim to the Strasbourg Institute's preoccupation with radiotelegraphy at the time.

Braun next turned his attention to the mixtures of iron filings and vaseline whose behavior under the influence of rapid electrical oscillations he had investigated in 1902. He came to the conclusion that they resembled crystals that exhibit optical birefringence and lead to a double image of objects viewed through them. Mandelstam and Brandes did indeed prove that such mixtures exhibited "magnetic birefringence."[83]

To confirm the identity of electromagnetic waves and light waves, Braun undertook to carry over into the realm of optics a property of electromagnetic waves demonstrated by Heinrich Hertz. Hertz had found that harp-like gratings of wires blocked electromagnetic waves under certain conditions, but transmitted them when the grating was rotated through a certain angle. In this experiment, the distance between the wires of the grating must be small compared with the wavelengths of the impinging electromagnetic waves.

It was not feasible to fabricate such a grating for light waves. It would have required wire thicknesses and spacings between $\frac{1}{10,000}$ and $\frac{1}{100,000}$ millimeters. But Braun knew that metal gratings much finer than any fabricated by a mechanic are obtained when a thin wire is placed on top of a glass plate and exposed to a current large enough to vaporize the wire. The vaporized metal explodes and is deposited on the glass plate in the form of an extremely fine grating perpendicular to the wire. By means of such a grating, Braun succeeded in demonstrating Hertz's phenomenon for light waves (see figure XVI in appendix B).

Braun considered his paper on this subject, "Hertz's Grating Experiment in the Visible-Light Range," so important that he submitted it to the Academy of Sciences in Berlin. It was to receive worldwide attention. "For the physicist," commented Braun in his paper, "there is nothing surprising in it, for the proof was sure to come sooner or later. But even if a phenomenon never before witnessed is awaited with complete assurance, there is nevertheless an element of surprise when it is finally observed."[87]

As counterpart of his proof that a phenomenon known from electromagnetic-wave experiments but unknown in optics must also occur in optics, Braun undertook an experiment in the reverse direction as well. He demonstrated the analog of an optical phenomenon, birefringence, in the realm of electromagnetic waves. He constructed an artificial dielectric in which the molecular dimensions relevant to optical phenomena were magnified a million times. He erected columns of bricks, arranged in several giant gratings, one behind the other (see figure XV in appendix B). As the number of layers of brick columns was increased, birefringence could be observed. Yet Braun had to break off the experiment before arriving at the optimum thickness, for practical reasons. "I estimate that I would have had to construct 11 or 12 such gratings. I was satisfied with 10; in optical terms that would correspond to a tiny crystal a few thousandths of a millimeter thick, but in electrical terms the configuration was 2.5 meters thick, weighed 80 cwt [hundredweights], and represented a cost of more than 200 marks."[83] There probably would have been no lack of bricks for the missing grating; indeed the Parliament of Alsace-

Lorraine had been quite generous ever since Braun had casually remarked, in a "wireless" lecture before the deputies, "Since the means for further research are lacking, I expect I shall be taking longer and longer walks." The real problem was that the weight of the bricks was too much for the laboratory floor. In 1887 Braun had spoken in Tübingen about "The Limits of Microscope Viewing," which were reached when the size of the object to be observed became comparable to the wavelength of the light with which it was observed. Now Braun speculated that the application of Hertz's grating experiment to optics might enable one to go beyond those limits.

Hertz had found that the wire grating blocked or transmitted electromagnetic waves according to the angle through which the grating was rotated, provided the distance between the grating wires was small compared to the wavelength. Braun's gratings fulfilled the same conditions for optical waves. The distance between adjacent elements was so small that they no longer could be resolved even with the finest optical microscope, precisely because the distance was small compared with the wavelength of the visible light. But as Braun's grating was rotated and its appearance changed from light to dark, the same phenomenon manifested itself; and the color of the transmitted light made it possible to calculate the separate microscopic grating gaps to an accuracy of 0.00005 millimeter.

Braun also speculated about biological applications. "Let us assume that a similarly fine grating structure exists in organic tissue such as muscle or plant fibers, perhaps in the form of tiny channels. If they could be filled with metal, the specimen would act optically as a Hertz grating."

During most of 1904 Braun sought practical applications of his discovery. Each day he spent hours preparing and examining thin slices of pinewood and nettle fibers soaked in solutions of gold and silver salts. He corresponded about the phenomenon with the Leipzig botanist H. Ambronn, whose earlier observations Braun had interpreted in the above fashion. He even thought of utilizing fossil materials that had become impregnated with iron over the ages. Specimens provided by his Strasbourg colleague Wilhelm Benecke enabled him to observe phenomena "from which one may draw conclusions about their original submicroscopic structure—a singular sensation."[88] These observations were described in a long paper of thirty-nine pages, "On Metallic-Grating Polarization, Especially Its Application in the Interpretation of Microscopic Specimens."[88] There were thirty-one colored plates showing his observations under the microscope. But before the scientific community could become familiar with Braun's method—Ambronn in Leipzig was particularly active in its development—Braun's thoughts had turned to x-ray microscopy. The limitations of light microscopy were ultimately overcome a quarter of a century later by the electron microscope. Braun had a part in that as well: the electron microscope is essentially based on his cathode-ray oscilloscope.

In 1905 Braun attempted a real *tour de force*: the creation of an artificial dielectric that would exhibit birefringence at optical frequencies. His report on some preliminary experiments, "Some Observations Relative to Artificial Birefringence," does not make it clear how the artificial dielectrics are to be constructed. He reported only that "the experiments had not brought him as close to his goal as he had hoped."[89]

In 1905 Braun reverted to the analogous phenomenon, natural birefringence, in a report on a particularly beautiful experiment in which a kaleidoscope of "vivid colors followed each other as the specimen was rotated."[90] The experiments were done with an exotic substance, tabasheer, which his friend Schär had found in the drug collection at the pharmaceutical institute. Braun was looking for birefringence in any stratified substance he could obtain. "There is unconscious poetry in the idea that when we move about in a thinly wooded stand of fir trees, in a hall of columns, or even around certain houses right here in Strasbourg, we are in an electrically birefringent medium, an artificial crystal," he wrote.[90] By means of an ingenious experiment, Braun demonstrated that tabasheer was a genuine birefringent grating—tiny columns of silicic acid separated by thin layers of air. "When the air is displaced by a liquid optically indistinguishable from the silicic acid, the birefringence disappears; it returns when the liquid evaporates—a game that may be repeated over and over again."[90]

The final word on this series of experiments was a paper published in the *Annalen der Physik* on 4 July 1905, to summarize what had been learned about the "Mechanism of Galvanic Pulveriza-tion."[91] Efforts to produce the most perfect pulverized gratings possible had even led to a doctoral dissertation, "On the Pulver-ization of Dynamically Heated Metals," by Gustav Äckerlein, an alumnus of Braun's *Thomasschule* in Leipzig, who had become Braun's private assistant in 1899 at the age of 20 and remained with him for fifteen years. Braun's paper describes how too strong a discharge is capable of pulverizing not only the wire but also the underlying glass plate. Vaporization of brass wires yielded layers of zinc (which has a lower melting point than copper) on the inside, surrounded by very fine, bright-red strips of copper. Even a piece of a rather expensive tantalum wire provided by Siemens & Halske, with a melting point of 2,200°C, was pulver-ized, which proved that extraordinarily high temperatures were reached in a matter of microseconds.

A curious episode of those years was the correction of what Braun later called "an almost incredible mistake." For nearly eight years it had been assumed that the wavelengths of the Marconi transmitter were determined by the configuration originally pro-posed by Hertz and accordingly were very short; the antenna was regarded merely as a convenient means for radiating the energy. Even Braun ascribed the higher efficiency of his transmitter only partly to a better configuration, and in greater part to the use of the "long" waves generated by a tuned circuit. As better wave-meters came into use it was found that the Marconi transmitter also generated "long" waves; the entire transmitter, including the antenna-to-ground circuit, acted as an oscillating system. A curious situation now arose. Since Braun had often referred to his own "long" waves, and since it now appeared that the original Marconi transmitter produced these same waves, it was easy to conclude that there really was no difference between the two systems. Braun sought to refute this "erroneous interpretation" in a scientific ar-ticle,[78] as well as in a popular weekly, *Die Woche*.[79] "In 1898 I deliberately used wavelengths of the same order of magnitude as those Marconi had employed inadvertently," he wrote.[75] To Fed-dersen he wrote in 1909, "It is so easy to forget later how matters stood in the scientific world at the time the discovery was made."[104]

In 1903, after five years of unsuccessful experimentation, Braun finally realized the idea, mentioned in his "energy circuit" patent of 1898, of increasing the transmitted energy by the almost simultaneous discharge of several oscillators into the antenna circuit (figure III in appendix B). In a 1904 paper on "Methods for the Amplification of Transmitter Energy in Telegraphy," he reported that he had harnessed the "usually capricious sparks" to make them wholly reliable; that is, the oscillating circuits could be discharged into the antenna at millisecond intervals.

The next logical step would have been for Telefunken to try to exploit the new circuit to improve its competitive position. It now became apparent, as Zenneck later put it, "how disadvantageous it is for someone who wants to implement technological advances to be on the outside of the industrial establishment and thus be unable to exert the requisite pressure for the testing and realization of his ideas."[132] Things were very slow getting started. Not until a year later was Braun able to report that Telefunken was working on the technical implementation of his "energy circuit."

In that respect his circuit shared the fate of another one of his inventions, the crystal detector. As mentioned previously, Braun had rescued it from obscurity when radiotelegraphy began to rely increasingly on audible reception, beginning about 1901. Braun-Siemens had hesitated to introduce a crystal detector because Siemens wanted to continue its production of the Köpsel spring-contact earphone receiver. After the formation of Telefunken in 1903, AEG-Slaby-Arco at first insisted on the use of their own liquid-detector earphone receiver.[293] In this way, although the advantages of the crystal detector were quite obvious, four years were wasted before it was introduced—a costly mistake. In 1905 the vacuum-tube rectifier was introduced in America and Telefunken sales there dropped sharply. Only then did Berlin wake up and quickly introduce a crystal detector, a simple and inexpensive device that was to dominate the field for years.

In February 1905 Braun lectured at the prestigious Institute of Naval Architects in Berlin on "New Methods and Goals of Wireless Telegraphy." His distinguished audience included the Kaiser. It was the first time His Majesty had been briefed on the new field by someone other than Adolf Slaby. A paper summarizing this vivid and entertaining lecture ultimately appeared in the institute's yearbook.[85] We also have something close to a verbatim transcript, prepared by Braun's 17-year-old daughter Hildegard, who as Loyal Traveling Companion and Private Secretary pro tem had a chance to observe the Kaiser's reactions at close range. Braun started with an analogy:

When a church bell is rung, we know the bell ringer makes it easy for himself: he draws on the pull rope, but always at the right intervals. He gives a little pull and then waits until the bell has passed through an entire oscillation; then he gives another pull at the right time. All his pulls add up until the sum of the individual pulls causes large movements of the heavy bell.

Now the spark—which has received the signal honor of having the whole thing named after it—is a thankless customer. On occasion I have compared it to Saturn, who ate his own sons. We can't increase its strength arbitrarily. We are led then to a simple conclusion: If we have already chosen a draft horse from the strongest existing breed, then to get more work done we shall simply have to use two horses instead of one. You don't need a professor for that sort of advice. (Laughter.)

But if I am to harness two horses, I must make sure they don't pull in opposite directions, else I should be better off with one; and if I take four horses and one pulls forward, the second backward, the third to the right, and the fourth to the left, I shall have four horses to feed and still I shall be unable to budge from the spot. So the trick is to train the four horses to pull together. . . . (The Kaiser laughs.)

So the spark must be set off with a precision of one hundredth of one millionth of a second. Easy to say, but consider that one hundredth of one millionth of a second is to one second as a second is to—I figured this out beforehand, else I shouldn't know it by heart (laughter)—three years. . . . I tried to make sure in every way that the sparks would set in at exactly the right moment, but it could not be done; I came close to saying, 'I can't do anything with these sparks, they are even more stubborn that I am.' (General amusement.)

But I would not let go and went back to try just once more, and this time it worked. . . . To begin with the circuits were so tightly coupled that when I drew energy from one circuit it was immediately replenished from the next. They acted like two ideal brothers—when one runs up debts, the other pays them. . . . (His Majesty laughs and shakes his head as if to negate this.)

But now the relation is transformed into that of a college student and his father. When one circuit runs up debts the other still pays them, but only after a certain phase delay. (General amusement.)

At this point Braun demonstrated his "energy circuit" in action. The printed text states, "Discharges a meter long and as thick as an arm were drawn from a coil." Hildegard noted that "It made a hideous noise." Braun introduced next the subject of directional transmission:

Now imagine the following scene at a radiotelegraphic station. Someone comes in. 'Your wish, Sir?—'Want to send a telegram.'— 'How far?'—170 km!'—The attendant motions with his hand: 'Done!'

Another person comes in. 'Your wish?'—'Telegram!'—'How far?'—'750 km!' The attendant calls to the back: 'Pull the first three stops!'

Third person. 'Your pleasure?—*I wish to telegraph to New York!*' [In English in the original.]—*'Well*—pull out all stops!' (Laughter, especially the Kaiser.)

'And you, young lady?'—'Me—only 100 km, but please, only in one direction!' (Loud laughter; the Kaiser is much amused.)

'I am terribly sorry, madam, that's something we can't do yet!' (Continuing laughter; His Majesty laughs and slaps his leg.)

The pretty lady's wish—I assume she was pretty, else the wireless attendant wouldn't have been so polite—her wish brings me to our next topic: In short, what are the prospects for achieving directional telegraphy? (The Kaiser draws up his chair as if to get closer to the speaker.)

His final remarks, greeted with long and loud applause from William II and the audience, were "Today electrical waves are used to bridge large sections of the earth. Soon they will allow us to view the world of microscopic beings." The concluding remark of the presiding officer, the Grand Duke of Oldenburg, put an end to the idea held by some of the country's leaders that all significant research in German physics was being done in Berlin.

Soon thereafter the Prussian minister of education offered Braun a chair in the capital.[123] What an accomplishment to follow Kundt and Kohlrausch, his predecessors at Strasbourg, to Berlin—the greatest honor that could be conferred on a German professor at the time! But Braun declined.

He did not like the rarified air that surrounded the court in

Berlin. Only the prospect of seeing his brother Adolf again over-came the considerable distaste that any trip to Berlin caused him. When the minister asked him whether he might now allow himself to be persuaded to accept the position in Berlin, Braun replied, according to his son's later account,

Where I am, I enjoy the comforts of a small university. I need spend only a moderate amount of time on examinations. In Berlin I should need more assistants to make up for time lost in exam-inations. In Strasbourg I have a large official residence right in the institute. I can work into the night if I want to, and then be in my bedroom in two minutes. In Berlin, I am told, one needs a car and a chauffeur on account of the long distances. On weekends I often make excursions into the country with one or two of my children. In Strasbourg we can take a third-class railroad carriage and be in the Black Forest or in the Vosges Mountains in an hour. To reach the mountains from Berlin means an overnight trip and—if we are to get any rest—a first-class sleeper. So I must ask myself: Can Berlin offer me the same conditions of time and freedom for my work and time and freedom for my own life in an environment I love?"[320]

It could not, and Ferdinand Braun remained in the Alsatian capital.

The University of Strasbourg rewarded its physicist's loyalty by electing him *Rektor* for the 1905-1906 academic year. The in-auguration of a *Rektor,* who is addressed as "Your Magnificence," is an important occasion. For his inaugural lecture, given on 1 May 1905, Braun spoke "On Wireless Telegraphy and Recent Physics Research." He offered his audience a glimpse into the coming atomic age. He showed how his own most recent inves-tigations on oscillations had penetrated into a field where "solids" no longer appeared quite so solid. He traced the discovery of radioactivity by Antoine-Henri Becquerel and radium by Pierre and Marie Curie and then moved on to atomic physics, where the limit of the divisibility of matter was no longer the atom but particles at least two thousand times smaller.

Braun noted that "Radium is warmer than its surroundings—it gives off energy as it disintegrates. The energy produced by one gram of radium would be sufficient to run a 30-horsepower au-tomobile traveling eight hours a day for a week, provided—as we would want it to in any case—it remained at rest on Sundays. We know of no other source of energy so highly concentrated. Its utilization might enable us to fly like the birds. . . ."

What was to be learned from the most recent developments of physical science? Braun's answer was widely disseminated because his inaugural lecture appeared in leading newspapers. It was a great scientist's preface to the new age:

During the past few years there has been an attempt to construct an opposition between pure and applied science. Pure science is indebted to applied science for its good repute, which justifies the sacrifices made for its sake. But before knowledge can be applied, it must first exist. And so applied science owes its very being to pure science.

It is no longer possible to construct a priori the new pathways of science. Almost always it can be shown that really new problems require new methods for their solution. That atoms are not the smallest particles—that they are not really *atomoi*—has long been a *conviction*. Yet all known methods of research had been exhausted; we could force no further knowledge out of nature by levers and screws. Yet she yielded new information freely to those who learned through painstaking research how to utilize fortunate coincidences and implied relations.

The problem of the old philosopher, the goal of the alchemists across the centuries, the dream of our youth—to transform one element into another—all that, if we are not altogether mistaken, we have now achieved. And so you who are students are not entering an exhausted land but one that has just been opened up. May you have success in civilizing it![86]

In trying to resolve the problem of directional transmission, Braun pioneered in the development of phased antennas, in which an array of antenna elements is excited in a prescribed sequence. The experiments, again performed on the flat Strasbourg drill ground, were evidently conducted with the cooperation of army authorities, which mounted a sentry over the three wooden sheds housing the apparatus.[124]

The first antenna configuration had been derived from the parabolic mirror of a searchlight. An array of vertical antenna elements, mounted on poles forming a cylindrical parabola, formed a mirror-like reflector that would turn the isotropic (nondirectional) radiation of an exciting element at the antenna's focus into a directional beam. Braun next hit upon the idea of simplifying this antenna curtain by reducing the number of the vertical elements to three and exciting them from a common transmitter in such a way that the signal arrived first at the central element and then, after a tiny phase delay, at the two outer ones—which was the electrical equivalent of setting them up in a geometric pattern. Braun arranged the delay circuits, which were derived from earlier experiments by Mandelstam and Papalexi, so that the direction of the resulting beam could be varied at will.[94] The method proved to be somewhat cumbersome,[95] especially for tactical military use. (The motivation for these experiments was evidently the desire to deny an enemy the chance of intercepting messages by making tactical transmissions highly directional. The development of the other major use for directional antennas—pinpointing a ship's position at sea by zeroing in on two land-based transmitters of known location and finding the intersect—proceeded along somewhat different lines.[362]) A variant of Braun's circuit ultimately found application in the design of radio transmitters for overseas traffic.

The "working university" of Strasbourg differed from other German universities. There was not much beer drinking or dueling; most of the student societies had only a few members.[204] Only outside support kept some of them alive, for instance, the student corps Palatia, which was supported by its sister chapter Teutonia in Marburg. The Teutons sent one of their members, a medical student named Hermann Kehl, to Strasbourg at the beginning of the 1905–1906 winter semester to shore up the outlying chapter's affairs. Kehl reported that the *Rektor,* Prof. Ferdinand Braun, was present at the first get-together of the new semester, delighted to meet a genuine Teuton from his old university. The *Rektor* wore the blue Teuton cap, "of the old style," reported Kehl, and participated "with youthful energy" in the ceremonies.[327]

Another festive occasion for His Magnificence was the traditional St. Nicholas Day *Rektor's* Ball, one of the high points of Strasbourg's winter season. The preparations occupied his family for some time, especially his daughters Hildegard and Erika, who took charge of the arrangements. The ball was one of the many good parties given by the Braun family, which as far back as the Tübingen days had made their hospitality famous. Only Zenneck, customarily the "master of ceremonies," was missing. The previous summer he had moved to the University of Danzig as an associate professor.

Zenneck had been sent a formal invitation from His Magnificence and had formally replied that he was coming.[135] The letter of acceptance was received as a jest. Yet it turned out that Braun had trained his student not only as a good physicist but as a practical joker worthy of himself, for in the midst of the ball Zenneck suddenly appeared. He had taken a train in Danzig and had sat up for twenty-four hours across the whole of Germany just so he could cap the jest of his written acceptance by the even greater surprise of his actual appearance.

Braun was to act again as Zenneck's patron before the vacation ended. The Technical University of Braunschweig was looking for a full professor of physics; Zenneck had been suggested for the post, and Braun had been asked if he could recommend him. Braun advised Zenneck to stop in Braunschweig on this way back to Danzig to introduce himself. Zenneck made the detour and a little later, at the age of 35, was appointed to his first professor's chair there.

Toward the end of 1905 it was diagnosed that Braun had rectal cancer. He was fortunate, though, in that a surgical technique had recently been developed to deal with the disease. Moreover, the most qualified person to perform the operation was Prof. Kraske, chief surgeon of the university clinic at Freiburg im Breisgau, not far from Strasbourg. The operation took place in January 1906. Four months later, when *Prorektor* Harry Breslau substituted for the *Rektor* in giving the annual report, he began his speech with the news that Braun's recovery was "proceeding most happily," a statement that made the great lecture hall ring with prolonged applause.[203]

For a time, Braun had to rest. The task of concluding the directional-antenna experiments then being conducted under military auspices was given to Dr. Papalexi who, Braun wrote, acquitted himself "with care and tireless effort."[94] The diligence with which Papalexi worked on the military grounds was attributed by some, even those close to Braun, as due to something besides scientific interest. Rumors went around that Papalexi was a spy. As evidence, it was pointed out that he seemed to have more money to spend than most of his fellow students and that he made frequent trips home. As a matter of fact, all military installations, not only the radio shacks, were guarded so carefully that any real espionage by Papalexi seems out of the question. At that time Strasbourg had the largest garrison in the Reich. The reason for the double guards at the radio shack was purely military busywork; the apparatus inside was the same as that supplied by Telefunken to the British army, the French Coast Guard, schools of military engineering in Russia, or naval stations in China. For his pocket money and trips home, Papalexi was indebted only to his father, a wealthy Crimean landowner and winegrower.

The control of wireless telegraphy by private companies led to situations that would have been the despair of a present-day counterespionage agent. For example, Braun's first report on a practical method of phase-delayed transmission, the basis for the experiments carried out on a drill ground sealed off by German army sentries, appeared in the *Comptes rendus* of the French Academy a year before its publication in Germany.[79] A more detailed report, including experimental descriptions, appeared in London in *The Electrician* in 1906 under the title "A Directed Wireless Telegraphy."[94] Fifteen months elapsed before an abbreviated version of this article appeared in German in the *Jahrbuch der drahtlosen Telegraphie*.[95] Research and commerce were, in short, much more free in those days than they are today. Companies were interested in secrecy only until their patents had been granted. Thereafter practical achievements were available to all paying customers on an equal footing. This degree of freedom has long since disappeared, even in democratic countries.

In 1907 Amalie Braun presented to her husband a dog of the Leonberger breed, who was promptly named Leo and remained a faithful companion for many years (figure 28). Leo accompanied his master everywhere, and the staff at the university could soon set their watches by Leo's prompt arrival at the lecture hall to await his midday walk. One day his master went beyond the usual hour and Leo became impatient. He put his great paw on the latch, pushed open the door, and, tail wagging, strolled into the hall. The students were delighted. "Na, Leo," his master greeted him, "since when have you become interested in physics?"

In the winter of 1907–1908 three of the Kaiser's six sons—

Figure 28
Braun and Leo.

Princes August Wilhelm, Waldemar, and Joachim—entered the
University of Strasbourg. Another royal student was Friedrich
August, Crown Prince of Saxony. When his father, the King of
Saxony, visited the university, Count Wedel, the imperial governor of the district, presented all professors to the king. Braun
drew a glimmer of recognition: "Ah! So you're the man who put
up all those wires in Dresden!" Although it was a great honor for
the university to have three princelings among its students, pro-
tocol sometimes caused problems. For example, at the *Rektor's*
banquet in 1908, Count von Wedel could not be invited together
with the Kaiser's sons because no one could decide who should
have the place of honor at the dinner table.[204]

Prince Joachim of Prussia asked Braun to give him a private
survey course on radiotelegraphy—a costly undertaking, since the
same demonstration experiments had to be set up as for a regular
course. The results scarcely justified the effort, though the prince
must have thought otherwise since Braun was granted an imperial

decoration at the end of the course. When he showed it to Rolf, his mechanic, Rolf pulled *his* reward from his pocket—a gold tie pin with the letter J and a crown above it. "You know what?" said Braun. "Let's swap!"[326]

Another unusual student to whom Braun taught physics during the years 1905–1908 rewarded him not with a decoration but with a letter that is a unique memorial to the kindness and affection that Braun bestowed upon each one of his students. This was an older student, a doctor of philosophy and theology and *Privatdozent* in theology at the university. Moreover, he was director of the Thomas Foundation and a preacher in the Lutheran church of St. Nicholas. In spite of all that, he had agreed to accept a call from a French Protestant missionary society to become a physician in French Equatorial Africa. His name was Albert Schweitzer.

Before he could embark on the twelve-semester medical course, Schweitzer had to obtain government permission to become a student at the same university where he taught—an unprecedented case. The 30-year-old Schweitzer was received with some amusement by his young classmates.[209] But Braun treated the extraordinary medical freshman with friendly understanding and consideration.[124]

Besides being a medical student, Schweitzer continued to teach and to preach. Only his iron constitution (he lived to be 90) enabled him to cope with the long hours. Sometimes he worked long into the night in the physics institute's workshops. With the knowledgeable assistance of the mechanic Meyer he fabricated all sorts of objects that he meant to take along to Africa. On at least one occasion the hour grew so late that Meyer would not let Schweitzer go home and provided a cot for him in the institute.[124]

Finally, in 1908, Schweitzer presented himself for the *Physicum,* the first interim examination for medical students. A later letter from Schweitzer states that Braun "treated me very gently" in that examination.[303] Who knows but Schweitzer might have become a physicist if the call to do good among "the least of my brethren" had not found such a strong echo in the heart of this Alsatian pastor's son. Indeed Schweitzer became so deeply absorbed in scientific research that he remembered the examination only at the last moment; like many another student before and since, he had to cram for a few days and nights from notes that were surreptitiously circulated and contained the questions most often asked and the answers expected by the several professors.[209] Braun's own favorite questions were as well known as the anecdotes with which he interspersed his lectures. His son Konrad reported that "some of [his stories] became so well worn that the students were apt to start chuckling before Father had quite come to the point. It did not seem to bother him."

Braun's convalescence after his operation meant not only a strict diet but also a reduction in his research output. No publications stem from the period between 1906 and 1910, though his many young assistants carried on with the development of radiotelegraphy. Among them were the two sons of the Bonn surgeon Otto Madelung, scions of a family of professors. Erwin Madelung, later professor of theoretical physics in Frankfurt, spent two semesters studying physics at Strasbourg, selected a project derived from the work of Varley and Brandes on a magnetic detector of electromagnetic waves, and then took his project to Göttingen where he became the first of the Braun school to obtain his doctorate elsewhere. Georg Madelung later became the scientific director of the German Aeronautical Experiment Station and professor at the Technical Universities in Berlin and in Stuttgart.[156]

The connection between high-frequency physics and aeronautics, which continues into the space age, was personified in Max Dieckmann, another alumnus of the St. Thomas school in Leipzig. Dieckmann studied in Strasbourg in 1906 and 1907; it was he who undertook, with Gustav Glage, to patent the previously mentioned "television" application of Braun's cathode-ray oscilloscope in 1906. He obtained his doctorate in 1907 with a dissertation on "Periodic Determination of Oscillations in Condenser Circuits," a phase-delay measurement technique deriving from the drill-ground experiments on directional transmission. Dieckmann became an expert in radiotelegraphy and aeronautics in the pioneering days of Count Ferdinand von Zeppelin (1839–1917) and later, from 1937 until 1945, served as director of the German Research Institute for Airborne Telecommunications in Oberpfaffenhofen near Munich.[156]

After Zenneck left in 1904, Brandes became first assistant and Mandelstam, second assistant. Brandes left after obtaining his doctorate in 1906 with a dissertation on "Damping and Energy Efficiency of Several Transmission Configurations in Radiotelegraphy," and Mandelstam became first assistant and later *Privatdozent*. Gustav Äckerlein became second assistant. This team remained together until the outbreak of World War I.

Among other assistants were Georges Rempp (1905), Alois Wuest (1906), Max Dieckmann (1906), and Gustav Glage (1908). Glage and Wuest later became *Gymnasium* teachers. All these assistants, except Wuest, wrote dissertations in high-frequency physics. Wuest wrote his dissertation on a subject that may have harked back to Zenneck's past interest in zoology: "On the Velocity of Propagation of Wave Motions and Single Pulses in Membranes Bounded by Water on One or Both Sides, with Special Consideration of Conditions in the Ear."

During his life Braun must have covered thousands of kilometers on foot: in the Rhön region near his home, in the vicinity of university towns, back and forth through the Vosges Mountains and the Black Forest, in Italy, on several Mediterranean islands, even in the Sahara. Best of all he liked to wander in the high mountains. He spent part of almost every one of his vacation months in the Alps. He enjoyed the trip, interspersed with short rail journeys, across the Alps from Voralberg to Trieste. At other times it was an extensive series of short walks to explore a particular region, for instance the Tyrol. Often he attempted real mountain climbs. He was a steady patron of the guide Tauscher in Oberstdorf. Braun's children regularly accompanied him on these excursions, which often ended with visits to the museums and theaters of Munich.

Braun's goal was to experience nature through leisurely observation. An animal near the trail, a rare plant, a beautiful rock, a view of a wooded valley, all would make him linger. When he was only 25, he wrote in *Der junge Mathematiker und Naturforscher*, "The more cultivated the mind, the more it will appreciate nature. That is why a landscape affects various observers differently, according to the development of their minds and temperaments."[6]

His appreciation of nature was no mere romanticism. The sharp spike of a nettle, a well-formed piece of gravel, the play of clouds, glaciers, and water all pleased him because he saw them as manifestations of natural laws: the orderly arrangement of cells, the manifestation of mechanical forces, the temperature dependence and states of water. It was as if he saw, within the first romantic picture of nature, "a second vision," one that edifies equally the mind and the spirit of man. The countless watercolor sketches that Braun brought home from his tours underlined what he had written many years before: "We cannot agree with those who believe that a deeper insight into nature destroys the charm of her beauty and interest."[6]

During his walks Braun maintained a slow but steady pace. If others rushed ahead, he would say, "Let them by; we'll get there before these foxes!"[311] He was rarely wrong. When skiing became popular at the beginning of the twentieth century, the 50-year-old Braun did not feel up to this new mode of winter locomotion; but he had a pair of Indian snowshoes made. After a few trials he gave that up, for he discovered that it was easier to tramp along in the trails left by skiers.[320]

On his Mediterranean trips he was usually accompanied by friends such as Johannes Thiele, a chemistry professor who had come to Strasbourg from the Netherlands in 1902, or the aviation pioneers Count Zeppelin and Hugo Eckener, with whom he had become acquainted through radiotelegraphy.[321] It was from one of these Mediterranean trips that Braun brought back the previously mentioned mosquito netting to Strasbourg, that "mosquito swamp." The netting did not last long. During one of his absences in Berlin Leo crawled into the bed and with his great paws reduced the netting to shreds.

A story about one of Ferdinand Braun's trips was published by one of his former students in the New York German-language newspaper *Staatszeitung* in 1924:

It happened at the Old Post Inn in Stuben near the Arlberg, whose landlady was famous for her cuisine. During one meal a heavy thunderstorm came up and so darkened the dining room that the electric lights had to be turned on. But the light would not work properly: it was bright one moment, quite dim the next. Asked the cause, Braun replied, "Haven't you ever heard of alternating current?" But when the light variations continued according to a certain rhythm he grew thoughtful, wondering whether the electrical discharges of the thunderstorm might not be somehow related to the variations in electric-light intensity. His son Konrad was dispatched to determine the source of the current. After a short while Konrad came back and reported with a laugh that he had discovered the cause of the variations. The electrical current at the inn was generated by a small waterwheel; but now a large milk bucket had fallen into the millrace, and whenever the bucket was being filled, water was diverted from the wheel and the generation of electricity slowed down. As soon as the bucket was thrown back by the water, it emptied, the wheel ran faster, and the lights became bright. The bucket was removed and the landlady's excellent meal could be finished under a steady light. Braun said good-humoredly, 'Thank God this difficult problem is solved, else it would have haunted me all summer!'[312]

In 1910 Braun leased a weekend house in one of the most beautiful parts of the Vosges Mountains, in Thannenkirch. One of his neighbors there was a friend and colleague at the university, the ophthalmologist Stilling. Stilling's son Otto provides another glimpse of Braun's personal life:

Unlike many scientists Braun was no ordinary laboratory toiler. Nature was no mere research object to him but an object for loving contemplation and absorption. I once noticed that when Braun and his colleague, Johannes Thiele, were both guests in my parents' villa in the Vosges. Thiele comported himself throughout as if he were at the lectern, forever lecturing. During their walks he continued to talk uninterruptedly, even when going uphill, whereas Braun said very little and probably paid more attention to the singing of the birds than to the scholarly outpouring of his colleague.[328]

By 1908 "Braun telegraphy" was ten years old and Telefunken, five. Both offspring had long since outgrown their father, yet he continued to watch their growth carefully. He made contributions to their technical development with suggestions and sometimes research of his own. Franke at Siemens wrote later, "His efforts were directed purely at furthering progress, and his manner was so warm that he became a friend and a valued advisor of all of us."[118]

The principal new discovery of Telefunken's first five years was that the mutual effects between a tuned circuit and an antenna inductively coupled to it were far more complex than had been assumed. There were mismatches: the antenna did not radiate all the energy supplied by the tuned circuit, but reflected some of it back into the circuit, so that energy was lost. This trouble was traced in part to the spark discharge, which was too long. The high-intensity spurt of energy generated when the spark was first struck was successfully coupled into the antenna, but the spark then persisted and conducted some of the energy back into the circuit. The problem was to extinguish the spark as soon as the tuned circuit had transferred its high initial energy to the antenna.

Braun worked on this problem in Strasbourg with a Swedish engineer, Ragnar Rendahl, a member of the young Telefunken team who had been temporarily assigned to him. The idea was to use the results of some of the old vaporization experiments to find a way to introduce a thin wire into the circuit so that it would be vaporized at the right moment and thus extinguish the spark. But this approach proved unsuccessful; the vaporized metal remained conducting for too long a time.

Rendahl solved the problem by means of a mercury spark gap in vacuum, but Telefunken could not get all the engineering difficulties out of the way in time to compete with another method, the rotary spark gap that Marconi had introduced in the meantime. In the end both methods were supplanted by the "quenched spark" developed by Max Wien in 1908. In this method the usual pair of spheres forming the spark gap was replaced by closely spaced parallel plates. This circuit yielded such rapid deelectrification in the spark gap that the spark was extinguished almost immediately after it had been struck, so that the antenna radiated a long, slowly decaying oscillation. Rendahl went back to Telefunken as chief of development and joined Arco in perfecting the so-called singing arc method, which produced a type of arc discharge modulated at audio frequencies, so that various transmitters could be distinguished by the pitch of the audible note they produced.[363]

The quenched spark effectively ended the era of the old, deafening spark transmitters and of Braun's circuit, on which so much effort had been expended. His later circuits were functionally quite similar to the quenched-spark transmitter. He evolved a system comprising a single tuned circuit whose spark gap was divided into a series of fifteen to twenty smaller gaps. Arco later acknowledged the similarity:

With such short gap length, quenched-spark phenomena doubtless occurred. Peaks of efficiency were occasionally achieved, probably as a result of hitting accidentally on the critical coupling, but were wrongly interpreted because the quenching process was not understood.[114]

Thus Telefunken came to depend on Braun's institute in Strasbourg for ideas, development, and personnel, providing in return suggestions, apparatus, and support. Gustav Glage, in his dissertation on "Experimental Investigations of the Resonant Inductor," thanked not only his teachers Braun and Mandelstam, but also the Telefunken company, whose support evidently extended even to graphics: in the published version of his dissertation Glage, the son of a minor railroad official of modest means, presented his results in several colors on expensive paper.

8
YEARS OF HONORS (1909–1918)

Coming into his lecture hall one day in November 1909, Braun was greeted by applause in the traditional manner of German universities—the stamping of feet—so loud and prolonged that his wife, in another part of the building, dispatched Babette Gütle, the nurse who had joined the household after Braun's illness, to find out what was going on.[329]

Nothing was going on. Braun took his usual place behind the lectern. He did not stop the students or ask the reason for the unusual demonstration. If anything he looked somewhat helpless and embarrassed. Finally the stamping subsided. Surely he would say something now? But he said nothing. He cleared his throat and began the lecture as on any other day. Was it possible that he knew nothing? Had he not seen the papers? But surely he must know, else he would have at least asked the reason for the applause. He did know. The Royal Swedish Academy of Sciences, the newspapers reported, had decided to award the Nobel Prize for physics for 1909 jointly to Marconi and to Braun, "in recognition of their contributions to the development of wireless telegraphy." But why was he silent? Why did he not thank the students for their ovation?

The next day brought the same routine: applause, quiet, beginning of the lecture. But this time one of the students rushed out at the end of the lecture, posted himself in front of the physics institute with a camera, and awaited the usual appearance of the professor and his dog for their usual walk. When they came out, the student motioned to the professor to stand still while he pressed the rubber bulb of his camera. Braun stopped obediently, then continued on his way past a group of applauding students.[329]

Not until the third day did Braun acknowledge yet another ovation:

Thank you, ladies and gentlemen, for all these demonstrations. They were evidently occasioned by the news reported from Stockholm, which had not been as yet officially communicated to me. Now it has, so that I can scarcely deny. . . .[329]

The rest was lost in a new and still louder storm of applause.

The Nobel Prize was still a fairly new institution in 1909. It was established by the last will of the Swedish chemist and industrialist Alfred Nobel, who in 1867 had invented dynamite and developed a worldwide business from its exploitation. Before he died, he set aside 31 million crowns of his wealth for a foundation with a most humane goal: Each year the interest was to be equally divided among the five persons "who have in that year most benefited mankind."

The Nobel Foundation was established in 1900. The first prizes were awarded in 1901, for physics, chemistry, physiology or medicine, literature, and peace. Nobel Prizes for physics had gone to Röntgen, Zeemann and H. A. Lorentz, H. A. Becquerel and Marie and Pierre Curie, Lord Rayleigh, Lenard, J. J. Thomson, Michelson, and Gabriel Lippmann. All these scientists had participated to some degree in the transformation of classical physics into the physics of the atomic age. The prize for 1909, divided between Marconi and Braun, was the first bestowed for more "practical" research, the importance of which even the man in the street could understand. Next to the awards to Röntgen and to the Curies, the 1909 award aroused the most public interest.

The joint award of the 1909 Nobel prize in physics to Marconi and Braun is an example of the awards committee's recognition of both the practical as well as the theoretical contribution. There was no better way for it to express its conviction that the development of radiotelegraphy owed as much to inventors as to academically trained scientists, the former as initiators, often erring yet making positive advances, the latter as elaborators of a body of knowledge accumulated over generations.

On 10 December 1909, the President of the Royal Swedish

Academy of Sciences, Hans Hildebrand, in a speech introducing the two recipients of the physics prize (see appendix A), differentiated their accomplishments: "A man was needed who was able to grasp the potentialities of the enterprise and who could overcome all the various difficulties that stood in the way of the practical realization of the idea. The carrying out of this great task was reserved for Guglielmo Marconi. . . . The development of a great invention seldom occurs through one individual man, and many forces have contributed to the remarkable results now achieved. Marconi's original system had its weak points. . . . It is due above all to the inspired work of Prof. Braun that this unsatisfactory state of affairs was overcome."

The five Swedish physicists who oversaw the prize attended to the details of the award: on the seal of Marconi's award certificate is a drawing of a ship with a "Marconi antenna" spreading out like a fan between the masts; on Braun's, a symbolic circuit diagram of the "Braun transmitter" (figure 29).

The two men met for the first time on Thursday morning, 9 December 1909, in the hall of the Grand Hotel in Stockholm.[319] Marconi was 35, Braun 59. They knew each other only from photographs, but they recognized each other at once. There had been much speculation about what might happen at this meeting. Would the Marvelous Marconi, so proud and unapproachable, who disdained his competitors in radiotelegraphy, snub the little professor from the provincial German university? Was it true that Marconi had never even heard Braun's name until the day of the award?[364] Had Marconi really refused at first to share the prize?

All these rumors were untrue, and the actual meeting of the two former rivals was pleasant and friendly. There was no reason they should not have been friends. Braun had never contested Marconi's priority in radiotelegraphy; on the contrary, he had often publicly praised Marconi's work. On his part, Marconi had never denied that he had made use of Braun's "important invention for increasing range." The two men had never opposed each other personally. The conversation that followed their greeting dealt with the division of topics for their Nobel lectures. Marconi, who originally had wanted to call his lecture "Telegraphy through Air," decided to leave it untitled. Braun's original choice, "On Wireless

Figure 29
Braun's Nobel diploma; his circuit is in inset above his name.

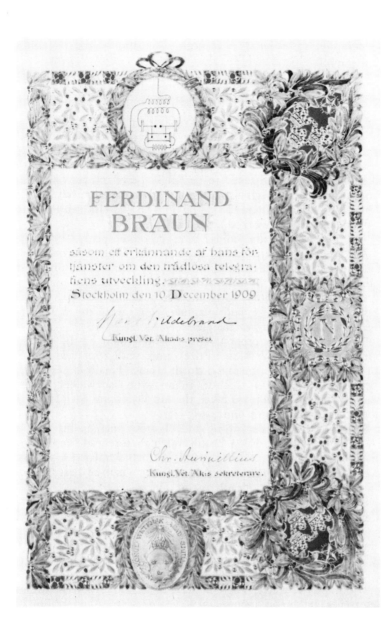

FERDINAND
BRAUN

såsom ett erkännande af hans för
tjänster om den trådlösa telegra-
fiens utveckling.
Stockholm den 10 December 1909.

Kungl. Vet. Akad:s preses.

Kungl. Vet. Ak:s sekreterare.

Telegraphy," became "Electrical Oscillations and Wireless Telegraphy."

A reporter for the Stockholm newspaper *Daghens Nyheter* described the arrival of the two Nobel Laureates:[319]

No one met Nobel Prizewinner Prof. Braun at the station when he arrived in Stockholm Thursday morning on the Berlin-Trelleborg train. The good professor doubtless would have declined to answer if anyone had dared approach to ask who he was. Most likely he would have asked to be left in peace until given Prof. Aurivillius's official signal that he might unmask himself. In the drafty lobby of the Grand Hotel I ran into his assistant, M. Pinchon [a Frenchman who worked at Telefunken].

"Why is Prof. Braun hiding?" I asked quite openly.

The Frenchman stammered: "Prof. Braun? Whom do you mean? I just got off the train from Berlin. Oh, may I introduce my wife, Mm. Pinchon? Now as to Prof. Braun. . . ." The charming lady interrupted at that point to rescue her husband: "Your Swedish trains run very quietly, and imagine, we came all the way from Berlin without having to change."

We regarded each other intently. Who would be the first to burst out laughing? A Nobel Prizewinner with a married couple for a bodyguard is something worth seeing. Finally I persuaded the pair to take me up to Prof. Braun, who had walked up the stairs while the others had taken the lift. A minute later I was in deep conversation with him.

Professor Braun is a strong and vigorous man in his sixties. He is not very tall, has a beard and high forehead, and wears glasses—all the traits of the typical German scholar. He was in high spirits, dampened only momentarily when he heard about the festivities that lay ahead.

He asked me to suppress my curiosity about his discoveries until after he had finished his substantial breakfast. Afterward, however, he vanished. Still, I thought I might get some information about him out of M. Pinchon. A short while later we ran into Prof. Braun again. We went up to the second floor with him, where, as an important guest, he had been furnished with two rooms overlooking the water. There he showed us publications about his researches and experiments and gave me some photographs of himself, which, however, did not please him very much. I must say he is partly right. The general features are there. But Prof. Braun is evidently a man of inner spirit, and that cannot be photographed. The photograph that he gave me was a snapshot taken by one of his students immediately after a lecture in Strasbourg.

The same reporter wrote as follows about Marconi:

Two o'clock in the afternoon. The small Marconi party emerged slowly, with a sort of British dignity, from the dining room of the Grand Hotel. Marconi's wife, the daughter of an Irish baron, came first; stately, elegant, a London lady to her fingertips. Then her sister, the Honourable Lilah O'Brien—the same carriage, the same pleasant face as her somewhat older sister, the same arch smile as she discovers the hundred-thousandth journalist on the lookout for her famous brother-in-law, Guglielmo Marconi. Both ladies wore hats and stylish traveling suits trimmed with fur.

Mr. Marconi and his secretary Solari walked slowly upstairs. They stopped at the landing. They lit cigarettes. The group moved slowly to the second floor.

I was asked to wait in an anteroom. A minute later the secretary motioned me in and I stood before the "Marvelous."

"I am in a great hurry," said Mr. Marconi hesitantly and regretfully.

"I can imagine; so am I."

"Yes, we're all in a hurry, aren't we?" Marconi sat down in a comfortable chair.

"Mr. Marconi, we have all heard much about you, so this conversation should not take long."

Marconi's blue eyes, somewhat tired, blinked skeptically. The ladies withdrew, but the secretary remained as witness. Marconi's face was motionless, stiff, smooth as ice. . . .

The conversation turned to wireless telegraphy. It had been asserted that Mr. Marconi had said he had never heard of Professor Braun even by name. The question was rejected as groundless.

"No expert who has studied or worked with wireless telegraphy can say that Professor Braun's name is unknown or foreign to him," said Marconi, "least of all I, who have utilized Professor Braun's important invention in my work to increase transmission range. After I had solved the problem of transmitting waves to a distance of 21 kilometers, Braun worked to increase the range to 100-300 kilometers and accomplished it by means of the so-called circuit of jars. I made use of that invention later."

Then Mr. Marconi spoke about having had the pleasure of a conversation with his famous neighbor next door. I now had the pleasure of telling Mr. Marconi about my conversation with Prof. Braun and how pleased he had been about the cordial greeting that his eminent competitor had extended to him.

Now came the coffee. The cigarette was finished. I stood up and listened while Mr. Marconi praised Stockholm, a city that roused his admiration despite the fog. It was lucky that the Nobel Prizewinner didn't see the city during one of its "blackouts."

The Nobel Prizes were awarded by King Gustav V of Sweden

on Friday, December 10, at 5 P.M. at the Swedish Academy of Music, before a bust of Alfred Nobel. The physics prize went to the Italian Guglielmo Marconi and the German Ferdinand Braun, the chemistry prize to the German Wilhelm Ostwald, the prize for physiology or medicine to the Swiss Theodore Kocher, and that for literature to the Swedish novelist Selma Lagerlöf.[297] The peace prize was awarded on the same day in Kristiania (now Oslo), the Norwegian capital, jointly to former Belgian prime minister Beernaert and French senator D'Estournelles de Constant. Each prize was worth 193,363 francs.

The Stockholm ceremony was followed by a banquet in the Grand Hotel, which lasted until midnight. In his toast Ferdinand Braun warmly acknowledged the difficult and often thankless task that fell to the Royal Academy of Sciences in making the prize awards, and ended with a salute to that world-famous learned society. Marconi praised "the impartiality of the Nobel Prize Committee, [thanks to which] the Nobel Prize for Physics is rightly considered everywhere as the highest reward within the reach of workers in that branch of science."[297]

The next morning Marconi and Braun addressed the Swedish Academy of Sciences and were elected corresponding members. As mentioned previously, the main title of Braun's talk was "Electrical Oscillations"; radiotelegraphy took up only a part of his talk. He also spoke of the rectifier effect and his grating experiments, with their use for viewing microscopic preparations. In his final remarks Braun expressed his great pleasure at the opportunity to approach his long-time personal goal of a "sparkless telegraphy" by means of Wien's highly efficient quenched spark gap (see appendix B).

Braun had planned to leave Stockholm that afternoon, but he stayed on because the King of Sweden had invited the Nobel laureates to another dinner at the palace. Afterward he and Marconi both accepted the invitation of the Nobel physics committee's chairman, Knut J. Ångström, to visit the University of Uppsala, the main center of Swedish physics research.

The reception that awaited the Nobel laureate Prof. Dr. Ferdinand Braun in Strasbourg turned into an unparalleled celebration. The physics institute was floodlit. There were fireworks. The student body gathered in front of the imperial palace and proceeded slowly in a torchlight parade of two columns, one on each side of the broad avenue, toward the university. The students formed a circle in front of the university buildings, sang the old student song *Gaudeamus igitur,* and at the last note threw their torches into the circle's center to make a large bonfire. It was the greatest honor that German students could bestow on a professor. Ferdinand Braun waved his thanks to them from the balcony.[321]

A few days later Braun was host at a banquet to which a hundred members of Strasbourg society, led by the imperial governor, Count Wedel, had been invited. The first cook of Sorg's, Strasbourg's best restaurant, had been engaged for the occasion. All was carefully organized. The young sons of institute mechanic Meyer, stationed on the stairs to be ready for any special errands, had an overall view:

Babette, Prof. Braun's nurse, brought us tidbits of all the wonderful food whose aromas pervaded the entire house. As soon as dinner and all the toasts were over, Babette told us to make ourselves scarce. Quiet as church mice we huddled in a dark corner. Then we heard the sound of violins from afar. A long train of guests in animated conversation appeared. They were preceded by three violinists. Behind them were Prof. Braun with Countess Wedel on his arm; then Count Wedel with Mrs. Braun, followed by all the other guests. The host led his gay and curious guests in a long polonaise through the brightly lit institute and showed them everything from lecture halls to attic, from rooms for storing apparatus to the workrooms for students.

In the meantime the great hall had been cleared, and after the long procession had reentered, dignified elder gentlemen and young assistants alike whirled their ladies in dance after dance. We Meyers had to help considerably in the cleanup. We very gladly undertook the pleasant task that Babette assigned to us of carrying the leftovers to our apartment. Mother didn't have to cook for a week.[124]

Radiotelegraphy was one of the last major technological developments in which old Europe led both the United States and Russia. It is true that Popov in Russia was among the earliest pioneers, and that a society for wireless telegraphy had been founded in America as early as 1885 on the basis of early experiments with induction telegraphy.[265] But the great impetus that brought success came from Maxwell and Hertz, from Marconi and Braun—all Europeans.

In 1902 the American engineer Lee de Forest founded his De Forest Wireless Telegraph Co., one of the first instances of native enterprise in the new field in America. Perhaps De Forest owed his success to an early recognition that wireless telegraphy need not be a mere replacement for wire telegraphy, but a novel means of communication in its own right, centered around audio reception and aiming toward radio telephony.[273] De Forest did not balk at adopting features of the Braun transmitter any more than Marconi, but he was dissatisfied with the coherer as a detector and put his inventive genius to work at that point in the circuit. A continuous, high-quality detector of the sort required for audio reception had been available since 1900: an ordinary gas flame—a phenomenon that Braun had investigated as early as his Leipzig days. De Forest next sought to replace the unwieldy gas flame by the electron beam of the vacuum tube. His work was later facilitated by Wehnelt's introduction of a reliable thermionic cathode in 1905.

A somewhat similar device had been proposed independently of De Forest by Robert von Lieben in Vienna. It was this device that had led Telefunken in Berlin to produce the Braun crystal detector. But before this development had had a chance to affect the market, De Forest had successfully encroached on the dominant position of the two great European concerns, Marconi's and Telefunken, in America.

The advent of the electron tube signaled the transfer of radio technology from university laboratories and workbenches to industrial factories throughout the world. The pioneering achievements of Marconi and Braun, and of American, British, French, and Russian researchers, were sufficiently advanced to be suited to commercial exploitation. Like all the others, Braun went on with research work at his institute in Strasbourg and pursued old and new ideas for the improvement and development of discoveries. But the tasks of transforming research into engineering practice, designing apparatus, and developing and adapting the design of transmitters and receivers to mass production were taken up by industry.

At Telefunken, Braun had become associated with a group whose capacity for further research and development was ideally combined with the practical and manufacturing experience of

Germany's leading electrical companies, AEG and Siemens. The close relations that Braun maintained with them after giving up his own company turned out to be mutually advantageous. From Count Arco and Hans Bredow at Telefunken he received the impluse for further basic research; his ideas and calculations were then turned over to engineers who knew how to convert them into forms that could be tested and then put into service in the development of radiotelegraphy.

The intent of the 1903 merger negotiations thus was realized, and Germany achieved a leading role in the worldwide development of radio. In 1911 the affluent Marconi companies had to negotiate an agreement with their German competitors on the basis of complete equality. Before this agreement the sharp rivalry between the two great concerns had led to some grotesque situations. The Marconi Co. trained telegraphers who were able to send and receive signals at incredible speeds. These operators were offered to Marconi customers as part of a new "secret telegraphy." Telefunken replied by offering instruments that permitted quick retuning: "With such instruments we could make it fairly difficult for an enemy trying to jam us by continually varying our wavelengths."[298]

The London physicist Nevil Maskelyne's contribution to the rivalry was a practical joke. During a lecture before the Society of Arts in London, Prof. John Ambrose Fleming wanted to demonstrate that Marconi sets could not be jammed by competing stations. To the surprise of the audience, "the tape of the Morse recording instruments kept registering, instead of the expected message, a single word repeated at regular intervals: Humbug. Maskelyne had provided interference with the signals of the giant Poldhu station by means of a small, almost toy-like set 'as an expression of his scientific search for clarity about the attainable limit of sharpness of tuning.' "[86]

After 1908 other solutions, partly outside the realm of Braun's circuits, came into prominence: the quenched spark, mechanical transmitters of various kinds, and triode amplifiers like those of Robert von Lieben and Lee de Forest, which supplanted the "gaseous unipolar detector" both in Europe and in America.

All these developments led to a considerable increase in the number of radiotelegraphic stations, especially in America, as well as to an increase in litigation. In 1911 the US government began to prepare a law for the licensing of radiotelegraphic systems. Only stations that had been in operation prior to a certain date would be exempt. This news led Ludwig Stollwerck in Germany to suggest that Telefunken ought to establish a branch station in America. With its own base on the American continent Telefunken no longer would have to depend on the cooperation of foreign companies. Because the projected deadline was fast approaching, quick action was necessary. Stollwerck at once instructed his American branch, Stollwerck Brothers, to look for a suitable plot of land. By the time Berlin had agreed in principle, he was in a position to announce that he had secured 60 acres of land on Long Island, favorably situated near to New York City. The requisite equipment was shipped immediately. A new firm, Atlantic Communication Company (Telefunken System of Wireless Telegraphy), with Ludwig Stollwerck as president and headquarters at 47-49 West Street in New York City, was charged with the installation and operation of the station. In this way the radiotelegraphy station Sayville on Long Island came into existence.

Braun's offspring—both in German radiotelegraphy and his private life—were growing up. In 1904 his oldest son, Siegfried, had just left to study law; his daughters Hildegard and Erika were, respectively, 16 and 11 years old. By the time his younger son Konrad was ready to leave home, Erika was 19 and Hildegard had been married for three years to a Strasbourg pharmacist, Dr. Paul Stadler.

After completing his secondary education in Tübingen, Konrad had just started his university science course when his maternal uncle, Conrad Bühler, offered him a position in his prosperous banking business in New York. Ferdinand Braun had no objection to such a sojourn in the United States; on the contrary, he realized that opportunities for a future engineer would be improved by increased knowledge of languages and of the world.[320] After Konrad had completed his year of compulsory military service, he left for New York in the autumn of 1912.

Meanwhile, in the Physics Institute in Strasbourg, 20-year-old Hermann Rohmann, who had joined the Braun school in 1906, received his doctorate in 1909 under Mandelstam with a dissertation on "Measurement of Capacity Variations by Means of Rapid Oscillations." With him as junior assistant, the trio Mandelstam-Äckerlein-Rohmann comprised Braun's supporting staff from 1908 until the outbreak of World War I.

In 1912 Dr. Richard Gans, who had received his doctorate under Braun in 1901, made a one-year visit as *Privatdozent* in Strasbourg. He was to become the most far-flung of Braun's students: he was appointed professor in La Plata in Argentina in 1912. In 1925 he returned to Germany and served as professor in Königsberg and Munich until 1947, as scientific advisor to AEG, and as member of the German National Research Council during World War II; he returned to La Plata in 1947 and died in Buenos Aires in 1954.

Nicholas Papalexi also became *Privatdozent* in 1911, for spectroscopy and x rays. The department thus consisted of Braun as full professor, Cohn and Mandelstam as associate professors, and the two *Privatdozenten*—the institute staff's largest size during the forty-seven-year history of the German University of Strasbourg.[203]

At the end of 1912, Ferdinand Braun's oldest brother Wunibald died after a full life of 73 years. The firm Hartmann & Braun AG, which he had helped to found and in whose management he had been joined by his sons, still remains as a testimonial to his successful entrepreneurial activity. The *Elektrotechnische Zeitschrift* said of Wunibald, "He was not a public figure; his activity was devoted mainly to the internal administration and sales organization of the enterprise, in which he exhibited an uncanny talent for selecting able representatives in all countries of the world. He was one of the first to introduce paid vacations of up to two weeks for his workers, according to length of service." Wunibald Braun evidently had the same skill in dealing with people that characterized his brother Ferdinand in the direction of the physics institute.

It all started quite innocently. In March 1913 a wire was strung from the tower of the Physics Institute in Strasbourg to a room in the mezzanine, where scientific measurements were made under the direction of Dr. Mandelstam. The objective was to determine distribution of the electromagnetic field originating from the transmitter on top of the Eiffel Tower in Paris, 440 kilometers away.

A month later the single wire had grown into a thicket of cables, wires, and ropes. A mass of people milled around in the courtyard of the physics institute: members of the "young team" led by Dr. Kaltenbach, who was in charge of these experiments; and assistants, *Dozenten*, and the two mechanics Rolf and Meyer, assigned to the task of hauling away at the pulley.[330] Dr. Papalexi was in overall charge.[99] From the window of the observation room he directed the work of Rolf and Meyer. A square wooden frame as large as a room and supporting thirty turns of antenna wire was first hoisted halfway up the building. Rolf and Meyer then took hold of two long ropes that dangled down into the courtyard and slowly turned the frame, to the left or the right, according to directions from Dr. Papalexi. The two mechanics joked with bystanders till Papalexi called from the window, "Hush! Prof. Braun can't hear a thing!"

The professor meanwhile sat inside the room, listening through earphones and indicating by hand motions how the frame ought to be turned. Papalexi transmitted the instructions to the mechanics below. Everyone had become dead still. The only sound was the distant beating of horses' hooves in the street on the other side of the building. Finally Braun took off the earphones. "There is an effect, but not distinct enough," he said. "We shall have to go higher."[99] That afternoon a new attraction engaged the attention of passersby: a giant wooden frame wrapped with wire on top of the tower of the physics institute, suspended from a tackle and slowly turning, first in one direction, then in the other.

Thus was the frame antenna tested in application for the first time, fourteen years after its invention in 1899; eventually it would lead to the replacement of large antennas, which could be strung on small ships only with difficulty, by antennas in which the same length of wire was compressed into a much smaller rectangular frame. "Since the experiments had to be made in the open and thus were evident to any passerby, a patent was applied for before the value of the invention had been determined experimentally."[99] Further experiments took place in Cuxhaven, and the application for a British patent was meant mainly as protection from "binocular espionage" from British ships passing the site.

The rectangular antennas were uncommonly directional, but

their sensitivity was so much diminished that the arrangement proved to be of no interest. At that time there was no way of amplifying the energy intercepted by the antenna. During the entire period of the stuggle for increased range, the highest possible antenna efficiency was essential. Even in the directional-antenna experiments on the Strasbourg drill ground between 1901 and 1906 it had been taken for granted that directionality was of no use if it was attained at the cost of the antenna efficiency obtainable in an ordinary isotropic antenna.

The advantages of the loop antenna could be utilized only after a method became available to amplify the incoming waves in the receiver.[132] The triode amplifier was not used in that way until about 1913. It was then that Braun came back to the loop antenna, an example of what Zenneck called the "fine instinct" that Braun always exhibited for what was essential at a given stage of development.[132]

In a yearbook article published on 6 December 1913, "On the Substitution of Closed Current Paths for Open Paths in Wireless Telegraphy," Braun presented his loop-antenna experiments as an accidental by-product of scientific measurements of the field of the Eiffel Tower transmitter.[99] The experiments Braun made in the early summer of 1913 resulted in a new patent, his last derived from radiotelegraphy. Although these patents extended to a number of industrialized countries, they provided only limited protection. At a time when both charlatans and scientists were working in radiotelegraphy, a company such as Telefunken would have needed more lawyers than engineers if it had wanted to bring suit for every infringement; and even that might not have resulted in redress.

The loop antenna enabled Braun to undertake the first field-strength measurement of a transmitter, namely, the one at the Eiffel Tower. Walter Ungerer, a visitor in Strasbourg at the time, reports,

One day my brother, Dr. Arnold Ungerer, took me to see the tower of the Strasbourg Physics Institute, to which a loop antenna had been affixed. We put on earphones and could clearly hear the Morse signals from the Eiffel Tower station by means of the well-known crystal detector. They were weather reports. The increase and decrease of signal strength as the antenna was rotated were clearly perceptible.[330]

Arnold Ungerer received his doctorate in 1914 with a dissertation on "Experimental Investigations of Braun's Dispersion Grating," which served as further confirmation that "the experiment was a translation of Hertz's grating experiment to the visible spectrum."[123] Another student of those years was Hans Riegger. His doctoral dissertation under Braun was on "Coupled Condenser Circuits with Very Short Spark Gaps"; he later became a section chief in the Siemens research laboratory. (Riegger was the inventor of the high-frequency condenser microphone. He died at the age of 43 in Berlin.[156])

On 13 September 1913, Ferdinand Braun set out with his daughter Hildegard on a grand tour of Italy. They spent four weeks visiting churches and museums in Florence, Rome, and Naples.[105] His diary records precise descriptions of countless paintings and sculptures. He sketched works of art and plans of buildings that particularly impressed him and bought a few things that pleased him. It was almost as if the 63-year-old Braun suspected that he was seeing these treasures for the last time. The trip was devoted not only to art but also recreation. Braun took care that he was served good food at his hotels. He painted watercolors and enjoyed the "marvelous glow of the moon from the terrace" of a hotel on Capri. At Monte Cavo near Rome, he wandered through "very beautiful woods of beeches, chestnuts, and oaks." He was taken with "a meat market where one could buy roasted meats." In front of a church in Rome he noted "many pious folk attending a mass out of doors."

We get a vivid impression of the scene of his departure from the Palace Hotel in Rome, with the staff lined up for tips: "Ten lire to the first waiter, two to the second, five for the head waiter, four for the porter, four for the liftboy, one for the night watchman, six for the house waiter, and ten for the chambermaid." And we may visualize Braun's ironic expression as he regarded the frescoes at the Villa Farnesina in Trastevere and noted, "The story of Psyche persecuted by the jealous Venus: Cupid is to loose one of his arrows at her so that she will fall in love with a simpleton (or be seduced by him); Cupid falls in love with her himself.

Venus turns to Zeus for help, etc. Cupid finally defends himself before the Olympians, Psyche appears, Mercury offers her the cup of immortality, they are married, and the Gods drink in celebration."[105]

The catastrophe toward which the world had been heading since 1911 occurred in 1914. Even the enlightened and liberal classes who had concluded that Europe's decades of almost uninterrupted peace would lead to an age without bloodshed, at least among civilized nations, came at last to realize in 1914, as Zenneck put it, that "a storm definitely lay ahead."[135] The hope that reason would prevail made it possible for life to go on as usual at the Physics Institute at Strasbourg as elsewhere. On January 7 Braun sent off the previously mentioned paper on the measurements of the field of the Eiffel Tower transmitter (it appeared in a yearbook for 1913),[100] and on February 7 he followed it up with a letter to the editor.[274] This letter (although its author could not have suspected it) was an indication of things to come. Its subject was the observation that the electrical dimensions of an antenna must be related to the size of the oscillator-circuit components if optimum performance is to be achieved.

Braun had noted the improvement that the "choice of a more favorable coil" would produce as early as the water-telegraphy experiments of 1898.[68] Some of the trips with the *Silvana* in 1899-1900 had been made with this goal in mind. But in his basic patent of 14 October 1898 Braun had not mentioned the necessity of the "choice of a more favorable coil." Not until his public lecture in Strasbourg in November 1900 did he speak of it, as reported by the Strasbourg *Post*: "The arrangement of Braun's transmitter requires—if it is to work at full efficiency—that the antenna and the primary circuit be tuned to each other."[242]

Guglielmo Marconi had stepped into this gap during the interim, in his British patent of 26 April 1900, No. 7777. He claimed the requirement for "internal resonance" in the transmitter as his invention. To be sure, Braun had described the priority situation in his article in *The Electrician* after Marconi had publicized the four-sevens patent, but even that had not put the matter to rest. Marconi had his patent, and even though Braun could prove that he had worked for years on the same subject prior to the date of application, he could not produce anything in print. In his letter of 7 February 1914 to the editor, Braun drew on Jonathan Zenneck for a written description of the Cuxhaven experiments. He also referred to the important part that the tuning had played during the 1899 negotiations with Lloyd's. After an interval of fifteen years the letter showed clearly how unfortunate it was that Braun had heeded his business associates' request for complete silence in 1898 and postponed "until much later my public description of my circuits and experiments."[274] But worse was yet to come.

Braun had been led to write the 1914 letter to the editor because Marconi had once again started pushing the four-sevens patent. One cannot say with certainty whether this episode was related to the worsening international situation. It had not been so long before that Prof. J. A. Fleming, Marconi's long-time collaborator, had paid tribute to the continuing use that radiotelegraphy had made of Braun's circuit and crystal detector.[250]

Braun's next trip to Berlin, on 3 March 1914, was a sad occasion. His third brother had died there at the age of 67. Dr. Adolf Braun had served his banking firm as president and later as director until his death. He had made a name for himself in financial circles and had served as chairman of a special committee on mortgage banking policy, a position of national importance. Slaby and Sir William Preece had died shortly before; the laws of mortality were inexorably exacting their toll.

In March 1914 Braun spent a week with his daughter Erika and the Mandelstam family in Menton on the French Riviera.[105] They tramped all around Cape St. Martin and made several day excursions in the vicinity. Again in Strasbourg, he performed some radiotelegraphy experiments with Count Zeppelin at Friedrichshafen on the north shore of Lake Constance. Papalexi, Ungerer, and the institute mechanic Rolf participated, as did Max Dieckmann, who in the meantime had become *Privatdozent* in Munich.

That was not the first time that Braun had become involved in aviation, then still based on balloons and dirigibles. As early as the winter of 1900-1901, in Braun's second public lecture on radiotelegraphy, he had spoken in Strasbourg before an "Upper Rhineland Aviation Society"; and the Royal Prussian Army's

Aviation Division had carefully monitored the Strasbourg drill-ground directional-transmission experiments between 1901 and 1903. The importance of radiotelegraphy for aviation was quite apparent to Braun. He had played an essential role in the development of the various special transmitter and receiver installations made available for this purpose with a "guaranteed range" of 25 or 100 kilometers.

At the conclusion of the Lake Constance experiments on Monday, 21 April 1914, Count Zeppelin invited Braun for a motorboat excursion from Friedrichshafen to the Insel Hotel in Constance. On their return to Friedrichshafen the two men had a last beer together in the Kurhotel. Braun noted in his diary that the 75-year-old Zeppelin seemed "quite vigorous, only a bit unsteady when he climbs in or out of the boat."[105]

In June 1914 Mandelstam and Papalexi were suddenly called to Russia for a "military exercise."[320] Of course they did not come back. It was the end of a fifteen-year collaboration that had been among the most fruitful in the new physics. Starting as assistants when they were both just 20, the two Russians had become successful independent researchers. Their diligence had helped the Braun school in Strasbourg to expand and the new science to become an independent branch of physics. Braun often expressed his thanks. At the end of his 1913 paper on the experiments with loop antennas, he wrote, "These experiments were made possible only by cooperative efforts."[100]

As mentioned previously, Mandelstam became a professor at the Polytechnic Institute in his native Odessa after the Bolshevik revolution in October 1917. In 1925 he became a professor at the University of Moscow. He died at the age of 65 in 1944. Papalexi, after giving up his postwar efforts to continue the Braun school in Germany, joined Mandelstam in Odessa and with him helped to found the Soviet Union's Central Radio Laboratory in Leningrad in 1923. He was also professor at the Technical University at Leningrad. He died in 1947.[318]

On 16 July 1914 an illustrious company assembled at *Spierlinsrain*, the summer residence of Adolf Friedrich Bader in Lahr. Forty-four Bader relatives were present to celebrate the 70th birthday of Amalie Braun's aunt, Frieda Maurer. A group picture was taken, which included the Nobel Laureate Ferdinand Braun and His Ex-

Figure 30
Ferdinand Braun, two weeks before the outbreak of World War I, in
family circle at Lahr: Prof. Carl Engler (left of Braun), Mrs. Frieda
Maurer, and his wife Amalie (right of Braun).

cellency Privy Councillor Carl Engler (figure 30).[323] A few days
later the First World War broke out.

FROM FERDINAND BRAUN'S DIARY

Friday, 31 July 1914: Lectures closed down; in the afternoon state
of war in Germany. The institute's wireless station quietly shut
down by two postal officials and two policemen. Shortly thereafter
a visit from garrison headquarters to requisition the building as
quarters for 700 men.

Saturday, August 1: Institute cleared to receive troops. Mother
to Appenweier at noon. Troops later reduced to 350 men.

Sunday, August 2: Straw delivered, used to fill mattresses.
Machine-gun troops. Hospital supplies brought into the office.

Monday, August 3: More machine-gun troops. [At this point
the diary contains a sketch of two men fishing in the River Ill,
with the caption "Strasbourg on the third day of mobilization."]

Tuesday, August 4: Siegfried reports for duty, is sent to Obernehnheim; Rohmann, to Hangenbieten.

Friday, August 7: In the morning French reported in Mulhausen.

Shots fired at people attempting to cut through wires in Broglie Platz [in Strasbourg].

Saturday, August 8: All day long machine-gun troops making preparations. At 9 P.M. they march off to Unternehnheim.

Sunday, August 9: Calm, clear day. Took walk to the Kehler Gate in the afternoon. To the veterinary with Leo. He was found blind in one eye. Mines laid outside London.

Monday, August 10: Visit to Privy Councillor Perrucke. Confidential information: the fight near Mulhausen went well for us.

Tuesday, August 12: Hot. Rohmann stops over at institute.

Wednesday, August 13: News station in the institute works, but still crosstalk between the two circuits. Corrected by coils. Afternoon at Thiele's.

Friday, August 14. Again hot. Good reception of Paris. The inner coil still poorly excited. Hussars move off. At 7 P.M. to Germania restaurant [the Bismarck Regulars].

Sunday, August 16. Mayor's posters urge people to leave. Rain, rain. Army volunteers move into the institute.[105]

Thus war disrupted Braun's life. The first day ended instruction at one of the world's best-known physics institutes. The second day tore his wife from him. And before the week was out his son and his assistant Hermann Rohmann (a physicist who had become the fiancé of his daughter Erika) had been called up and the war had been brought to his very doorstep.

On Tuesday, August 18, Braun and his daughter Erika heeded the appeal to leave Strasbourg, which was threatened by military operations.[105] They took a train to Tübingen; because of the many troop trains, the trip was complicated and took two days. Braun hoped to find help from friends and relatives; but even that quiet university town had been deeply disturbed. Before they left the railroad station, Braun's daughter was called to work as an x-ray assistant on a train full of wounded soldiers that was just pulling in. The 64-year-old refugee went by himself to look for a place to live. He found a "great paucity" of rentals, visited former Tübingen colleagues, and noted "misinterpretation of our mood in Alsace." Not until Sunday, August 23, did he succeed in finding quarters for himself and Erika.

At the end of this day, exhausted from the several days of apartment hunting, he wrote down the thoughts that had been agitating him since the outbreak of the war. He certainly did not suspect that the break-up of his family and the separation from his workplace would be permanent. He still thought of his trip to Tübingen as one might regard closing the shutters during a passing storm. Yet in fact everything, the entire edifice of his long, fruitful life, already had collapsed into ruin.

He was under no illusion about the role that his life's work, radiotelegraphy, would play in the war that had just started. As early as 1905, in his inaugural lecture as *Rektor*, he warned that "malevolent use could be made of wireless telegraphy" in wartime; that it could add a hitherto unknown horror to warfare. In the Russo-Japanese war, it had helped bring about the sinking of the Russian flagship *Petropavlovsk*. "Small Japanese cruisers lured it out of the harbor of Port Arthur. As soon as it had come out, the cruisers used wireless telegraphy to call Japanese torpedo boats standing by to the attack" wrote Braun. Such bravura feats of a new "wireless" warfare might charm the military, but they depressed the pioneer of radiotelegraphy.

The rest of the diary entry for 23 August 1914 deals with the recognition that the rise in the level of scientific education and of objective thinking had not made war among civilized nations impossible: "The dream of my generation in its youth—a dream that appeared closer and closer to realization after the year 1870—that war, at least among civilized nations, was an impossibility, *is a utopian dream*."

Science had not fulfilled its promise of leading the peoples of the world toward permanent peace through the understanding of natural laws. "This experience with the psychology of nations is like a scientific discovery that might be understood afterward, but that cannot be deduced a priori without hypotheses."

The months of "evacuation" in Tübingen passed quietly. Erika was so very busy at the university clinic that she could not take much care of her father. In September Frau Amalie therefore left Appenweier for Lahr, where she could be cared for by her relatives. That freed nurse Babette Gütle, who came to Tübingen with Leo.[105] In the 27-year-old Babette, who had been with the Brauns for nine years, Ferdinand Braun again had the experienced cook for the strict diet necessary to his health and in Leo the faithful companion of his walks. With letter writing, visits, and water-colors, the days passed.

Welcome news came that his son Konrad was well. At the

outbreak of war in Europe, Konrad had left America by the first Dutch liner, to report for military duty in Germany. The ship had been intercepted by a British gunboat, and the German passengers were interned in England. Konrad had been able to send a reassuring message home through an intermediary in London.[320]

As the front moved westward and Strasbourg passed out of immediate danger, Braun returned there on 19 October 1914. It was cold and foggy and a drizzling rain was falling as he arrived in Strasbourg in the evening, after another difficult journey that had been frequently interrupted by trains loaded with troops and wounded. The last kilometer, from Kehl across the Rhine, had to be covered on foot.[105]

The physics institute proved to have been severely used. The mechanics and those of their families who had remained behind had not been able to keep constant watch on everything. "Not a bottle of wine. Most annoying," noted Braun after a visit from his son Siegfried, who was on duty as mayor of the neighboring Alsatian hamlet of Obernehnheim and had pedaled on his bicycle to see his father on October 25. Braun noted in his diary on the next day, "I shall save my remaining cigars for frontline soldiers on leave."

Lectures were resumed after a few weeks, although the number of students was sadly depleted. The participants were recent high-school graduates who had not yet been called up, wounded men from the Strasbourg military hospitals, and members of the Strasbourg garrison.[207] For the first time since the outbreak of war, the great loop antenna at the top of the institute tower was placed back into operation. The connecting cables led into a room closed off from all unauthorized personnel. Army operators were using it to intercept messages from British ships in the Atlantic. The direction of the source, when compared with data from two similar stations in Cologne and on Germany's north coast, made it possible to determine the locations of all ships sailing to or from Britain. On the basis of these reports, German warships and, later, submarines were dispatched to operate against the British supply lines. It was the first large-scale radio triangulation in the history of warfare. The duty of directing it had been assigned to Hermann Rohmann.[105] There were closed doors also in the nearby Strasbourg Chemistry Institute.

On 17 December 1914 Braun, entered in the passenger list as Karl Braun from Thannenkirch, sailed on the Norwegian steamer *Bergensfjord* from Bergen in Norway to New York. A fellow passenger was Jonathan Zenneck, who had barely been given time to exchange his army officer's trunk for a plain trunk.[135]

One of Britain's first military operations after the outbreak of the war had been to cut the transatlantic cable between Germany and the still neutral United States. After that the newspapers of the most powerful neutral nation became largely dependent on French and British sources.

Britain understood that news is a weapon; hence a second attack was directed against the German colonial radio network. Thanks to the superiority of Braun's telegraphy, this network had been the first one ready for worldwide traffic. While the British network was still tied up in parliamentary debates, the distance from Nauen near Berlin to Kamina in German Togoland in West Africa (5,400 kilometers) had been bridged by 1913. The entire network was completed shortly before the outbreak of war. It extended to the Cameroons, Southwest Africa, East Africa, and, via a Dutch cable, to the German possessions in the South Pacific. When the war started, stations operated by Telefunken throughout the world were attacked and destroyed. Only one connection to the outside world remained to the Central Powers: the private radio station in Sayville on Long Island, which had been constructed in 1911 through the efforts of Ludwig Stollwerck as a counterstation for Telefunken. But Sayville was not prepared for war. It was intended only for coastal communications with arriving and departing ships. Not until a "singing" quenched-spark transmitter was installed there on 12 February 1914 was it possible to transmit messages to Nauen. But with a power output of only 35 kilowatts, the voice of Sayville was so weak that it was lost in the roar of the wireless traffic across the Atlantic during the inferno of war.

Immediately after the outbreak of war Stollwerck's Atlantic Communication Co. began to increase the Sayville transmitter's power. The first task was to improve the receiver. Within a few weeks of the interruption of the transatlantic cable in September 1914 German war communiqués again became available to American newspapers. They were received with skepticism. How deeply the American public had already been influenced can be gauged by the fact that it believed that the first German reports had been "invented in Sayville."[135]

The Allies now determined to interrupt this "overseas wireless cable" as well. First, British diplomats made representations, which led to no results. Next, the American Marconi Co.—a branch of the British parent firm—sued Atlantic Communication Co. for infringement of Marconi's four-sevens patent. The real objective was the shutdown of the Sayville station.

Berlin looked forward to the trial with equanimity.[114] It would be easy to prove that the internal tuning of the transmitter of the four-sevens patent had been a feature of the Telebraun transmitter before Marconi had availed himself of it. What was alarming was the news that Marconi was to appear in court in person "to validate his company's alleged rights with the authority of his personality."[114] Only one way remained to counter this threat. The testimony of the Nobel Laureate Marconi would have to be opposed by that of Nobel Laureate, Braun. Only then could the New York court be expected to arrive at an objective judgment.

Everyone concerned understood what that meant for Braun. At 64 years of age, in failing health, he would have to embark on a difficult winter journey through the British blockade; even if he arrived safely in America, he would probably have no chance to return until after the war. Breaking through the blockade meant risking capture by the enemy, imprisonment, and internment in enemy territory. These circumstances made it difficult for those at Telefunken who had to suggest the voyage to Braun. Arco wrote later,

Professor Braun was fully aware that this journey, particularly under existing conditions, quite likely would be the last act of his life and that he could scarcely expect to see his native land and his family again. His illness, which had started ten years before, had been temporarily arrested by surgery; but shortly before his departure fresh symptoms began to manifest themselves.[114]

On 29 November 1914, Braun received a telegram asking him to come to Berlin. There the facts were put before him. He immediately declared himself ready to undertake the trip. "The love he bore his life's work and his staunch patriotism led to his decision. Despite his poor health, he undertook the strenuous and dangerous journey."[114] At the beginning of December Ferdinand Braun bade farewell to his wife Amalie. It was to be a last goodbye. Even with the best hopes for a quick conclusion of the litigation and safe return by mid-1915, neither could count on seeing the other again. Amalie Braun, now 56, had found a quiet refuge after the excitement of Strasbourg's first war days in Achern in Baden, where she owned a house. But her health was no better than her husband's. Selflessly she did him a last service: she persuaded Babette Gütle to accompany him as nurse and housekeeper.[321]

Braun's farewell to Strasbourg was also difficult, especially since he was unable to tell the university, the institute, or even his coworkers the real reason for his trip.[303] To keep his important mission a secret the Prussian ministry of education had represented his trip to administrative and university authorities in Strasbourg as a professorial exchange. Most of those who were told of the "exchange" probably drew their own conclusions. Some guessed that Braun was "being sent to the United States on a diplomatic mission, with other well-known persons, to persuade America not to intervene in the war on the European continent."[124] No one had any idea about the struggle for Sayville.

Remaining behind in the Strasbourg Physics Institute were Cohn, Rohmann, Äckerlein, and a small number of students. After the German University of Strasbourg was shut down in 1918, Cohn became honorary professor of theoretical physics in Rostock and Freiburg until his retirement at the age of 68, in 1922. The Third Reich took away his professor's title in 1935. In 1939 he took refuge in Switzerland, where he died in 1944 at the age of 90.[156]

Rohmann became *Privatdozent* in Tübingen and Münster after the war and director of the physics laboratory of a mining company in Gelsenkirchen. He died in 1931 at the age of 45. Äckerlein became assistant in Frankfurt and later professor of physics at Freiberg, a school of mines in Saxony. He retired in 1950.

On 8 December 1914 Braun traveled with his nurse to Berlin. The last direct news from his family reached him there the next day: a second child had been born to his daughter Hildegard.[105] In Berlin, Zenneck and a Telefunken engineer, Thiel, joined them. Zenneck, who had been hauled out of the trenches on the Western front, had declared himself ready to accompany Braun as a witness to the early Cuxhaven experiments.

Of a last conversation Count Arco wrote, "The mood in which Braun, under foreboding circumstances, on a sad December evening, took leave of me will never leave my thoughts."[114] Then Braun, Babette, Zenneck, and Thiel left for Norway to board the steamer *Bergensfjord* of the Norwegian-American Line (figure 31). The little port of Bergen near the Arctic Circle had been chosen because the *Bergensfjord* sailed directly from there to New York and because embarkation in a larger neutral port meant running the risk of recognition and discovery of the entire enterprise.[135] As it was, no one on board identified the illustrious passengers. Their luck held throughout the three-week crossing. The boat was crowded, but because of the ceaseless December storms, few passengers trusted themselves out of their cabins. No one noticed what Zenneck discovered when he looked in his cabin mirror, that a strip of pale skin extended clear across the top of his forehead, unmistakably revealing he had been wearing a helmet.

Figure 31
Circuit sketches in Braun's notebook, made during wartime transatlantic crossing.

The ship's captain, K. S. Irgens, made every effort to avoid the British blockade. Zenneck recalled later,

The day after we left Bergen we tried to determine our course with the help of a watch compass that Prof. Braun had brought. According to the reading we were far north of the great-circle route to New York. At first we thought that our compass might have been affected by the iron smokestack, a supposition that turned out to be false. We were even more surprised a few days later to see, during a spell of clear weather and blue sea, a large island with high mountains covered with snow and glaciers. We asked the captain about the island. He replied, with a sly smile, that he had seen it too but did not know what it was.[135]

It was Iceland. Deviating from the usual direct course that led close to the northern tip of Scotland, the *Bergensfjord* had sailed far to the north in order to stay out of the blockade area. The captain meant to deliver the valuable "contraband" aboard his ship safely to New York. This cargo included not only the four Germans traveling on a secret mission, but also, of much greater interest to the Norwegian-American Line, two disassembled metal antenna towers intended for the Sayville transmitter, together with a new transmitter that would ensure reliable two-way transatlantic transmissions. Thiel was along, in fact, to direct the installation of the new transmitter. "*Bergensfjord* would have been a fat morsel for a British cruiser."[135]

A few days after Christmas the *Bergensfjord* docked in New York. The passengers went through the usual immigration formalities. For the first time in his life, Braun made the acquaintance of a quesionnaire.[135] Do you believe in anarchy?, Do you believe in bigamy? were questions to which he could answer no with a clear conscience.

Shortly after the arrival of Ferdinand Braun, another Braun arrived in the port of New York. It was his younger son Konrad. Through the good offices of American and British acquaintances of his uncle Conrad Bühler, Konrad had been released from British internment on Christmas Eve and was allowed to return to America. Shortly before landing in New York Konrad had received a wireless message aboard ship: FAMILY MEMBER MEETING BOAT. He assumed that meant his uncle—and sure enough, there he was on the pier. But who was the bearded gentleman next to him? Konrad Braun was puzzled until the ship drew nearer. He could hardly believe his eyes: the gentleman was his father! It was a highly emotional meeting.[320]

For a few days father, son, and housekeeper were guests of Conrad Bühler in his home in Greenwich, Connecticut. Then Braun rented furnished rooms on Manhattan's East Side, where he set up his own household, to be run by Babette (figure 32). The need to take up residence had become evident right away because of the slow progress of the litigation. The judge of the US District Court for the Eastern District of New York, before whom the case would be heard, seemed determined to come to a just judgment, which meant prolonged examination of witnesses; moreover, as Zenneck noted, "A patent case in America is of interest to all sorts of other people. Everybody participates vigorously: the lawyers, experts, professional people. . . . A patent trial between two prosperous companies is like a saucer of sugar water on which the most diverse insects can feed for months."[135]

Two roles were envisaged for Braun by Knight Brothers, the

Figure 32
Ferdinand Braun, in a photograph taken in New York (1915).

Atlantic Communication Company's New York attorneys. The first was to counter Marconi's personal appearance by the weight of his own personality. It soon turned out that this problem would not be a factor in the trial, since Marconi did not appear. The second role was to provide support for Zenneck, who was to testify on the Cuxhaven experiments of 1899-1900. Several boxes of the original equipment used at Cuxhaven had been brought along; meanwhile, it had been donated to the Deutsches Museum in Munich by the Cuxhaven authorities. But at the very beginning of the preliminary hearing, the opposing lawyers succeeded in obtaining a ruling to the effect that testimony regarding events that had taken place outside the United States was inadmissible.[135]

The entire mission to America appeared to have been in vain. But fortunately Knight Brothers persuaded the court to accept Zenneck as expert witness on behalf of Atlantic Communication Company, a decisive success. As to Braun, by April 1915 he realized that it was no longer necessary for him to remain in the United States. His trip had served its purpose without his court appearance. Marconi's lawyers yielded the field in Sayville to the American Telefunken Company and took a different tack. In the expectation that the Brooklyn court would not hand down a favorable ruling, they asked that the trial be postponed indefinitely. In the meantime, they filed suit against a small radiotelegraphy company on the other side of the American continent: Kilbourne and Clark Manufacturing Company in Seattle. They figured that this company would be in no position to sustain the effort and costs of an American patent suit. (The Brooklyn suit had cost Telefunken half a million marks up to the time of postponement.[135])

If the Marconi attorneys won in Seattle, they were expected to resume their suit in Brooklyn. Accordingly, Atlantic Communication Company lent Zenneck to Kilbourne and Clark. During a three-week court battle Zenneck succeeded in having the company acquitted on all major points. Zenneck was immediately put into action for the next litigation against Atlantic Communication Company, and it looked as though he would be kept busy for some time. Braun, however, could think about returning home now that the four-sevens attack had definitely been averted.[367]

On 17 May 1915, Braun applied to the German Embassy in Washington, asking that a "safe conduct" be obtained from the

British government through diplomatic channels.[105] His state of health and his advanced age led Braun to believe that his request would be granted. The British government replied promptly, on May 27, but without any definite guarantee. It merely stated that "persons not of military age and not engaged in non-neutral activities are generally not molested. . . . That might also apply in the case of Professor Braun." Braun was not reassured by this declaration. Instead, he decided to follow the advice of Conrad Bühler, to leave his flat during the approaching summer and take up residence in the Catskills, in a small German-American colony, Elka Park. He could continue his efforts to return home from there.[320]

The move to the summer residence was made by Hudson riverboat. Braun spent the next several months in the pleasant Catskills environment taking walks, sketching, and visiting friends in the vicinity. He paid an almost daily visit to his neighbor, the lame Dr. Hasslacher, to discuss world and war news, derived mainly from the German-language *Deutsche Staatszeitung*, which was published in New York. He was also a frequent visitor at the colony's small German club, where he often talked with the then famous journalist and Hearst correspondent Karl H. von Wiegand.

For the first time in his life, in Elka Park Braun lived in the country. He found the experience invigorating and refreshing. To assure a supply of fresh eggs, Babette Gütle in her practical way acquired a few chickens, with which the professor helped her. This first intensive encounter with agriculture made a great impression on him. He drew and painted pictures of the chickens as they went scratching about and pecking.[320] He dreamed of returning home and acquiring a small house in the country near Munich. He wanted it to have a view of the Alps and to be properly equipped with horse and carriage, so that he could reach both city and mountains comfortably.

But the war raged on and home was far away. One summer's day an excited neighbor came in to report that he had heard a telegram being sent by a man, evidently a British detective, asking to be relieved of keeping a watch over Prof. Braun, a task he found monotonous. Braun thought it amusing. "Let him come up," he said, "and move right in. We can use a handyman and gardener around here. And if the British want to pay for it, so much the better!"[320]

At the beginning of October 1915 Ferdinand Braun left the Catskills and moved back to the city. He rented a small furnished apartment for himself, Konrad, and Babette at 677 Vanderbilt Avenue in Brooklyn. He preferred Brooklyn to Manhattan because it was cheaper and because the broad, tree-lined avenues and the front yards reminded him of home. His apartment was not far from Prospect Park, where he could take his daily walk. When the weather was warm, he sat in the park and painted.

He realized that his return was growing more and more doubtful. At the end of the summer of 1915 Atlantic Communication Co. made another attempt to get him a safe conduct, again without success. On October 21 the company wrote,

Consul Hossenfelder has stated that no one is in any position to issue any guarantees, in spite of your age.
An important consideration is that the British have a special interest in keeping you, as a world-famous scientist, away from Germany. A possible internment would be very dangerous for your health. It seems advisable, therefore, that you forego a return trip for the time being. If you absolutely insist on it, however, Consul Hossenfelder will do everything in his power [to help].[105]

The continuing presence of a German Nobel Prizewinner in the United States awakened interest among the general public and sympathy in professional circles. Soon after their arrival in New York, the Institute of Radio Engineers gave a dinner for him and Zenneck "at which we were treated most cordially;"[135] Braun was elected a Fellow of the Institute. The Faculty Club of Columbia University also gave a banquet in honor of their European colleagues. Braun paid frequent visits to the New York office of Atlantic Communication Co.; to the Siemens and Telefunken representative, Dr. Karl Frank; and to patent attorney Jahnke in the Knight Brothers firm. He also dined occasionally with the director of the Atlantic Co., Hermann Boehme, with whom he had become very friendly.

In this period Braun became acquainted with several notable American physicists. By attending lectures given by the Institute of Radio Engineers he managed to keep in touch with the latest developments in radiotelegraphy. He was particularly interested to find, in the course of these meetings, that American high-frequency physicists "knew decidedly more about the properties of electron tubes and had more experience with them than was the rule in Germany."[135]

An indication of Braun's interest in the further development of high-frequency technology was his chapter on "Wireless Telegraphy" in the volume on physics in the series *Contemporary Culture,* which appeared in Berlin in 1915.[101] In his review, prepared in 1914, Braun presented a complete summary of research in wireless technology up to the most recent problems of the excitation of oscillations and of "wireless telephony." Concerning the latter, he wrote,

Transmission of speech is said to be quite clear. The distances achieved are very considerable. The Poulsen transmitter in California, with an antenna 90 meters high, claims a range of 550 kilometers. Gesellschaft für drahtlose Telegraphie [Telefunken] has transmitted speech from Nauen to Norddeich and, according to the most recent reports, with new transmitter arrangements they have perfected transmission from Nauen to Strasbourg and to Vienna. Newspaper reports were read for an hour. Sound level at the receiver station was more than adequate.[101]

From the drawings of wiring circuits and diagrams in his diaries and on bits of paper we know that Braun remained actively involved in wireless technology during his last years.[105] At the same time he was interested in other branches of physics. Even in a foreign land he could not live without science.

As in the early years at Tübingen, when unsuitable conditions at the institute for experimental work had forced him to work primarily on theory, so now Braun again turned to theory in America. Professor Alfred Goldsmith offered him laboratory facilities at Columbia University, but Braun did not feel well enough to accept this offer and limited his experiments to paper. When visitors arrived one day and asked Babette Gütle whether he was busy, she replied, "He is sitting upstairs, drawing little crosses and stars!" Braun was much amused when told his housekeeper's description of his activity.[320]

From this period we have 364 manuscript pages of tables, calculations, drafts of scientific publications, drawings, and fair copies in German and English on the following subjects:

nature of the liquid state,

absolute magnitude and law of molecular forces,

electrostenolysis: an attempt to explain the phenomenon,

investigations of thin silver layers obtained by the reduction of thin halogen silver mirrors,

extremely thin dichroic metal films,

the nature of photochemical (particularly photographic) processes.

The papers and drafts date from the summer of 1915 until the spring of 1917. (In 1962 Braun's son Konrad gave them to the Deutsches Museum in Munich.) None ever appeared in print. One possible reason was the strong propaganda in favor of Britain. From the fall of 1916 on, an increasing hostility toward Germany became noticeable in the United States. Thus in 1915 Zenneck was able to deliver a scientific lecture at the Institute of Radio Engineers; but in early 1917, when a similar lecture was announced, a newspaper campaign forced him to cancel the talk.[135]

In the spring of 1916 the incision from his surgery began to give Braun trouble. He found a sympathetic and capable physician in Dr. Willy Meyer of the German Hospital (now Lenox Hill Hospital) in New York. Dr. Meyer treated the professor as a colleague and refused any fee.[320] The treatment necessitated a one-month stay in the hospital. Shortly after he was admitted, Nikola Tesla, who had learned of Braun's misfortune from the newspapers, sent him a basket of splendid tropical fruit. Braun had been honored to meet Tesla at the 1915 dinner of the Institute of Radio Engineers (see frontispiece) and had come to appreciate him, despite the mixture of genius and madness in Tesla's personality. (A month later Braun read in the newspaper that Tesla had been forced to move out of his rooms at the Waldorf Astoria because he had not paid his bills for months, which led him to wonder whether the fruit he had enjoyed had been part of Tesla's debts.)

The summer of 1916 was spent again in Elka Park. In America Braun lived on a salary from Telefunken, agreed upon before he had left Germany. A return trip home for Babette Gütle was provided in case of his death. He was extremely conscientious about his financial affairs. His son Konrad reported,

When I visited him at the end of the Summer of 1915 in Elka Park he explained to me that if he had been in Germany, he would have spent a certain amount for his vacation. Since Telefunken was now paying his expenses, he had written to his bank to credit the company in Germany with this amount.[320]

During his illness German-Americans unknown to him wrote to wish him a speedy recovery. The editor-in-chief of the Swedish newspaper *Svenska Dagbladet* asked for a statement on how the war would affect international cooperation in cultural fields. A publishing firm, Göschen, asked him for a paper on "wireless telegraphy." Although he was far from home, Braun remained in contact with his country and with the world at large.[104]

In August 1916 the German Embassy informed him that the British and French governments had once more refused his application for a safe conduct. The British Ambassador, Cecil Spring Rice, wrote to the US Secretary of State, Robert Lansing, who acted as intermediary, "His Majesty's Government can see no reasons to make special arrangements for Dr. Braun. He must undertake the journey at his own risk." Very little hope remained that Braun would see his country again before the end of the war.

In his second autumn in New York, Ferdinand Braun undertook his last project, a manuscript entitled "Physics for Women."[113] The lack of a reasonable science education in girls' schools was something that had always disturbed him, and he missed no chance to try to change this state of affairs. "Every housewife ought to study physics and chemistry for a few months; then she would do things a lot differently," he said.[321]

At the age of 14, in his essay on water, young Ferdinand had noted, "Lentils do not cook well in hard water because a thin film of calcium forms on them and prevents the water from penetrating, or at least weakens its effect. But if the water is first boiled, it serves the purpose very well."[1] And when he was 25, in the person of "the father" in his book *Der junge Mathematiker und Naturwissenschaftler*, he taught his "daughter" how to tell whether a material was genuine silk: "Touch a charged electroscope with it: the leaves of the electroscope should remain separated."[6] Lack of time had prevented him from embarking on a full-scale text until the end of 1916 when, with political events preventing him from doing research, he finally set about realizing his old idea of writing a "Physics for Women."

The extant manuscript consists of an introduction, a nearly complete first chapter, and notes for additional chapters. It was written in English, but a number of difficult expressions were left in German for later translation. He foresaw two audiences for this book: the German *Hausfrau*, who would read it for its practical hints, and the woman with a more profound interest in the problems of science. His goal was to lead the reader from her everyday experience to general phenomena.

The completed first chapter dealt with "The constitution of matter—or how to clean a glass." In lively fashion, Braun uses the example of dishwashing as an introduction to the world of molecules and the calculus of probability. Mathematically speaking, no material entity adhering to a glass or dish—a bit of leftover wine, bacteria, a trace of poison—can ever be completely rinsed away with certainty. "But let us suppose that after so-and-so many rinsings only one bacterium is left in the glass; then it *can* happen that this bacterium will end up in the rinse water. But neither is it impossible that an unfortunate accident will leave it in the water left over in the glass. . . . Only because matter is not infinitely divisible, because it consists of exceedingly small particles, which are called molecules and are similar to the bacteria, is it possible to obtain a liquid free of any other matter."

Only notes exist for the second chapter, "thermodynamics for women." From these it is clear that the introduction was going to be done by reference to fever and that illustrative examples were going to be drawn from the workings of a warm oven, the heat exchange in human clothing, and similar practical things.

The two last pages of notes for "Physics for Women" are also the last extant instances of Ferdinand Braun's handwriting. They contain thoughts about mankind's oldest questions: life after death and the creation of the world. Referring to the interest of old people in the possibility of life after death, he writes, "The answer, scientifically speaking, is neither yes nor no. Faith. I have formed no judgment of my own. (a) All peoples have such belief—this proves nothing, but it might be interpreted as favorable. All connections in nature have been first surmised, then investigated, and—often—discovered. (b) We arrive in this world with many capabilities. But even the most intelligent cannot guess what it really is like." As proof Braun refers to the discoveries of electricity and radioactivity, which could not have been anticipated any more than a language that a people has made for itself from a given number of spoken sounds.

Concerning the origin of the world, Braun has these notes: "It might be thought of this way: some single separate collisions in the perfect equilibrium of space. Radiation, that is, change from the outside. That's when 'time' begins. That is, 'creation' is synonymous with the origin of time. Time: there are concepts of time only where there is change and memory. (In dreamless sleep time does not exist for us—but it does for those external to us.)"

In April 1917 the United States declared war on Germany, as might have been anticipated from the deepening anti-German sentiment. For Ferdinand Braun this event meant a new hope that he might be allowed to return home with the diplomatic and consular officials, but nothing came of it. He accepted this last setback with equanimity. In view of the untapped and enormous resources of the United States, with which he had now become acquainted and whose effects on the outcome of war he was well able to estimate, he felt certain that the war could not last much longer.

Unlike Zenneck, who had been accused of being a "German spy" even before war had been declared and who was immediately interned when it came, Braun was totally untroubled by the authorities or by public hostility. His age and reputation were respected.[320] He spent the winter of 1916-1917 in Brooklyn again, in a furnished apartment on Second Avenue. In the summer of 1917 he went back to Elka Park. He wrote no more, but gave his full attention to the progress of the war.

News reached the United States only from Allied sources now. Telefunken's Sayville station had been taken over by the U.S. Navy on 9 July 1915 and operated for its German owners after suspicion arose that they were violating American neutrality by slipping in coded information about Allied shipping among commercial messages. On the entry of the United States into the war, the station was augmented and ultimately became part of the Allied transatlantic communications network.[367]

Apparently all Braun's efforts, including the dangerous wartime crossing, were coming to naught. He could no longer speak out as scientist and inventor on Germany's behalf. And yet the fact that radiotelegraphic connection with Germany so far had been maintained was an important gain from the German viewpoint. For two and a half years German news services had broken through the British blockade to America and so to the world, thus to some extent countering the effect of Allied propaganda. Also, some technical advances had been occasioned by the need for keeping in touch. Telefunken had strengthened its Nauen transmitter to such an extent that its range extended all over the world. A station had been erected under difficult circumstances in Chapultepec in neutral Mexico that relayed messages from Nauen throughout South and North America. Toward the end of 1917 signals from Nauen were received at Avanui in New Zealand, 18,000 kilometers away and almost exactly at the antipode. Radiotelegraphy had encircled the earth. What had seemed an unattainable achievement at the time of the Helgoland experiments had become a reality twenty years later.

Meanwhile the father of radiotelegraphy had to rely on old-fashioned mail service. But sending mail through neutral countries took weeks. It troubled him to remain for months in uncertainty about his family and friends. When he did get news, very little of it was of the kind that might counteract the increasing slowdown of his creativity. Friends in Strasbourg, Laband, Stilling, Hermann, Count of Solms-Laubach, all contemporaries, had died. Johannes Thiele was fatally poisoned during the development of poison gas in his Chemistry Institute in Strasbourg.[206]

How often had Braun and Thiele traveled to Bingen to enjoy a carafe of wine together at the Society for Catholic Citizens! All that was gone forever.[311] The pain could be only superficially relieved at the news that the Technical University of Vienna had awarded Braun an honorary doctor's degree on 20 June 1917, in recognition of his contributions to wireless telegraphy. This degree was the only honorary doctorate ever awarded to Braun.[303] Other honors of this sort might have come to him if the war had not put an end to all amicable international relations that normally lead to such gestures.

In the summer of 1917 came the news that the condition of Braun's wife Amalie had worsened greatly. Before his return to Brooklyn from the Catskills, Braun learned that she had died on 17 September in her house at Achern near her native town of Lahr. She was 58; they had been married for thirty-two years.

The summer of 1917 had taken a heavy toll on Braun. His face and figure had become haggard, his full head of hair had grown sparse, and his entire physical condition had deteriorated, yet he never lost his sense of humor. It was well that Konrad had found at 208 Midwood Avenue a small house in a quiet neighborhood where there were front yards all along the streets. Nearby Prospect Park provided a place for Braun's daily walk. He rested on a park bench, watched the playing children, and now and then engaged in his favorite pursuit, painting watercolors. At home he immersed himself in newspapers and magazines and occasionally received visitors.

Autumn passed, then Christmas and the New Year. February 1918 brought cold weather and snow. After each snowfall passersby could see Braun clearing the path in front of his house. That is what he was doing one morning when the mailman brought a letter from Dr. Willy Meyer in reply to a letter in which Braun enumerated recent health difficulties and some minor domestic annoyances because Babette Gütle had been disabled by a cold. "I was sorry to hear that you have a regular hospital at home," wrote Dr. Meyer. "But I am pleased to learn that you remain a tower of strength in all the misery. By all means continue to shovel snow and to hack away at the ice—only don't overdo it!"[105]

On that morning Ferdinand Braun slipped and fell on the icy steps of his little row house and broke his hip. An ambulance took him to the German hospital. The injury would not heal. In addition there was discomfort from the old wound of his operation of many years before. He suffered greatly. Home a few weeks later, he had to stay in bed most of the time, racked by pain. His still critical and passionate mind realized that he would not get up again from his sickbed. "Bab and my son do everything for me which they can read in my eyes. Without them, I would not know what to do," he wrote on 3 April 1918; these are his last surviving written words.

Ferdinand Braun died on 20 April 1918. He had thoroughly refuted the theme of his own high-school graduation essay, written in 1868, that life was "too short for us to expect too much from it."[158] The richness of his discoveries and inventions, some of them recognized only after his death, had elevated him far above the usual scientist who, according to the thesis of Braun's Berlin teacher Wilhelm Dove, could count himself lucky "if he had one truly important idea in a lifetime, unless he was another Newton."

"A pioneer of science and a pathfinding German inventor," read the death notice published by Telefunken, "and for us, moreover, a friend who enjoyed life with a youthful heart, a noble person with unique simplicity of thought and feeling. His eyes, now closed, ever looked into the future. His name and his work will live on." The news reached Germany on 26 April 1918, through diplomatic channels by way of Switzerland. Despite the turmoil of the last months of the war, obituaries appeared in many daily papers and professional journals. These eulogies reflected Braun's importance as a teacher, for many of them were written by former students active all over Germany as scientists, science reporters, or physicians. In foreign lands as well, he was remembered.

9
EPILOGUE

It had been Braun's last wish to be buried in his homeland. Since the war made that impossible, his remains were cremated. The memorial service was held at the Campbell Funeral Church in New York. A few days after his death an officer from the U.S. Navy came to inspect the home of the "enemy alien." He discharged his task most tactfully.[320] The historic greatness of the deceased was recognized across the lines drawn by war.

Konrad Ferdinand Braun took over his father's household and took charge of his scientific papers. He continued to employ Babette Gütle until 1930, when she married a German-American named Strunk. She died in 1949 in White Plains, New York. Konrad Braun died in Kingston, Rhode Island, on 9 September 1965, a few days before the German edition of this book was published.

An opportunity to return his father's ashes to Germany did not arise until 1921. Strasbourg, where Amalie was buried, meanwhile had become French and did not seem to Braun's relatives the proper place for the interment. Ferdinand Braun's brother Philipp therefore obtained permission for the urn to be placed in their parent's grave in the Old Cemetery in Fulda. A memorial service was held at the grave on Saturday, 4 June 1921, two days before Braun's 71st birthday.[314] Only a few persons attended because the service coincided with a rally called for a diocesan Catholic Day. A minor official represented the town, and one reporter, from the *Fuldaer Tageblatt*, was present. This was Adrian Meyer, a former student of Braun's and a member of the old Philomantic Society of Alsace-Lorraine, over which Braun had presided until 1914.

Besides Ferdinand Braun's children, his brother Philipp, and his sister Katharina, Jonathan Zenneck was also present along with a former student named Dr. Rohn, and several schoolmates and friends of Braun's youth. (After his release from American internment, Zenneck had resumed his position as professor at the Technical University of Munich, where he had been since 1913. In 1933 he became chairman of the board of trustees of the Deutsches Museum. For the museum's collections he had reproductions made of Braun's original apparatus, which had disappeared in the United States. Zenneck died on 8 April 1959, at the age of 88.)

Graveside speakers included the Protestant pastor of the city of Fulda, Superintendent Ruhl, and Philipp Braun. In his eulogy, Philipp said of his brother, "He remained a simple person who thought little of great honors. We must hope that the spirit of taking pleasure in one's work, so characteristic of him, will be preserved and increase in our land. If it does, we shall be able to look forward to better times."

Ferdinand Braun's last public letter had gone to the German-language New York *Staatszeitung* on 5 April 1915. It was a critical comment regarding an article that had appeared in the newspaper about Michael Pupin, professor of electromechanics at Columbia University.[105] Pupin, whom Braun had befriended, was the author of a theoretical analysis demonstrating that losses in telephone lines could be reduced by the introduction of coils. The "Pupin coils," inserted in the lines every few kilometers, were considered a great technical achievement. They had made the extension of telephone communications to hundreds of kilometers possible even before vacuum-tube amplifiers were introduced. The article had failed to acknowledge this contribution and Braun now wrote that "my sense of justice leads me to draw your attention to this point. . . . Professor Pupin has been long known not only for his scientific investigations, but also for their engineering applications. In view of the importance of the resulting practical achievement, it seems to me that your article fails to acknowledge sufficiently to whom we are indebted for them."

It seems almost as if Braun had anticipated his own fate in this letter. There cannot be many scientists whose work has seen such lively subsequent development while their names have almost been forgotten. To be sure, it is not a unique case, but it is an extreme one—so extreme that one is led to speculate on the sources of scientific fame. We know such fame is not invariably commensurate with achievement; too many extraneous factors can intervene. The difficulty of the subject is one of them: if a scientific contribution is in an esoteric field of no immediate practical significance, not much of a public echo can be expected. The greatest scientist America produced in the nineteenth century, Josiah Willard Gibbs, is scarcely known, doubtless because his main work was in such a field, statistical mechanics. Another adversity derives from working in a place that is not in the mainstream of scientific activity. Gibbs had the further disadvantage of living in what we should today call a less-developed country, the United States of 1839, and it was only by the greatest of luck that his work came to the world's attention at all. Unfavorable ambiance has a stunting effect on scientific recognition—we need only think of poor Gregor Mendel, working out the fundamentals of genetics in the obscurity of his Moravian monastery and receiving no credit at all until years after his death. Another extraneous factor is the absence of a champion, the appreciative contemporary who turns out a well-written biography, or sometimes an institution such as a commercial organization or even a government. Would Madame Curie's name be a household word if she had not had a daughter who wrote a best-selling biography of her mother? Perhaps Marconi would not be so well known if his name had not been perpetuated by a great firm; certainly no one would know much of Popov if the full weight of the Soviet propaganda apparatus had not been thrown behind the effort to establish the simplistic and easily grasped idea that he had "invented radio."

Small wonder, then, that Ferdinand Braun never became widely known, since *all* the factors mentioned above were against him. Some of his earliest contributions were in an esoteric field, thermodynamics. He worked at provincial universities away from the great centers of the scientific research establishment. In several cases, long delays intervened between his discoveries and their practical applications. And before now he has had few champions.

Among Braun's many contributions, five stand out: the rectifier effect (1874); the concept of free energy in thermodynamics (1878); the cathode-ray oscilloscope (1897); the indirectly coupled, tuned, directive system of radiotelegraphy (1898-1900); and magnetic compounds (1902). We thus have a scientist who improved on a theory of Kelvin and Helmholtz, who discovered important properties of semiconductor and magnetic materials, who invented the ubiquitous picture tube, and who made an advance over Marconi's first circuits significant enough to share the Nobel Prize with him—yet who has remained practically unknown. Why?

To begin with, Helmholtz adroitly revised his own theory before Braun's correction became generally known. Except for certain industrial uses (in the conversion of ac to dc), semiconductor rectifiers did not come into general use until the advent of broadcasting in the 1920s and were largely superseded by vacuum tubes in the 1930s, only to return as the only suitable microwave rectifiers in the 1940s and to give birth to transistors in the 1950s—all a long time after the original discovery of 1874.[365] The delay between his first compound magnetic core and the introduction of ferrites was even longer. As to radiotelegraphy, the development of vacuum-tube detectors, oscillators, and amplifiers obscured the advantages of Braun's circuits, and even though variants of them continued to be used for a long time, the original motivation for them—a stronger spark—had vanished. The invention with which his name is most closely associated, the cathode-ray oscilloscope or "Braun tube," remained familiar only to a limited number of scientists and engineers until the much later advent of television. By then the memory of its creator had faded even in Germany, as Zenneck noted in a 1947 commemoration of the device's 50th anniversary. "Despite the importance of the Braun tube," said Zenneck, "his name is almost forgotten among physicists."[133] But there are at least three other reasons.

First, Braun died far from home, in a country at war with his homeland, virtually cut off from his compatriots. The turbulent and restless times that followed left little leisure for such niceties as keeping up the memories of a vanished age. In other countries anti-German feelings ran high. The hope expressed by Count Arco in his 1918 obituary of Braun remained unfulfilled: "Once the light of peace shines again and science, mankind's greatest boon, can weave a band that will tie together the cultures of all nations, recognition of Braun's achievement will come wherever true science is at home."[114]

Second, Braun's base was washed away by the war when Alsace-Lorraine reverted to France and the German University of Strasbourg was disbanded, so that the place where his memory might have been preserved academically disappeared. The succeeding French University at Strasbourg took over the buildings and the equipment but cast off the scientific traditions. There is a bust and a plaque in the Strasbourg Physics Institute commemorating a scientist who worked there, but that scientist is not the Nobel Laureate Braun.

Finally, no biography of Braun appeared for nearly half a century after his death, until the original version of the present work. The obituaries and memoirs that did appear[114-134] were mostly quite short pieces; the most extensive of them, a long section in Zenneck's own memoirs,[135] has never been published. In his 1947 lecture Zenneck cited a letter from an American colleague, who wrote that Braun was "a forgotten genius at present. It is my hope that something might be written to do him justice."[133] We trust that the present book has fulfilled that hope.

APPENDIX A

Nobel Prize Presentation Speech by Hans Hildebrand, President of the Royal Swedish Academy of Sciences, 11 December 1909*

Your Majesty, Your Royal Highnesses, Ladies and Gentlemen.

Research in physics has provided us with many surprises. Discoveries that at first seemed to have but theoretical interest have often led to inventions of the greatest importance to the advancement of mankind. And if this holds good for physics in general, it is even more true in the case of research in the field of electricity.

The discoveries and inventions for which the Royal Academy of Sciences has decided to award this year's Nobel Prize for Physics also have their origin in purely theoretical work and study. Important and epoch-making, however, as these were in their particular fields, no one could have guessed at the start that they would lead to the practical applications witnessed later.

While we are, this evening, conferring Nobel's Prize upon two of the men who have contributed most to the development of wireless telegraphy, we must first register our admiration for those great research workers, now dead, who through their brilliant and gifted work in the fields of mathematical and experimental physics opened up the path to great practical applications. It was Faraday, with his unique penetrating power of mind, who first suspected a close connection between the phenomena of light and electricity, and it was Maxwell who transformed his bold concepts and thoughts into mathematical language, and finally, it was Hertz who through his classical experiments showed that the new ideas as to the nature of electricity and light had a real basis in fact. To

*Translations of appendixes A and B follow substantially those of *Nobel Lectures, 1901–1921*, Amsterdam: Elsevier (1967).

be sure, it was already well known before Hertz's time that a capacitor charged with electricity can under certain circumstances discharge itself oscillatorily, that is to say, by electrical currents passing to and fro. Hertz, however, was the first to demonstrate that the effects of these currents propagate themselves in space with the velocity of light, thereby producing a wave motion having all the distinguishing characteristics of light. This discovery—perhaps the greatest in the field of physics throughout the last half-century—was made in 1888. It forms the foundation not only for the modern science of electricity but also for wireless telegraphy.

But it was still a great step from laboratory trials in miniature, where the electrical waves could be traced over but a small number of meters, to the transmission of signals over great distances. A man was needed who was able to grasp the potentialities of the enterprise and who could overcome all the various difficulties that stood in the way of the practical realization of the idea. The carrying out of this great task was reserved for Guglielmo Marconi. Even when taking into account previous attempts at this work and the fact that the conditions and prerequisites for the feasibility of this enterprise were already given, the honor of the first trials is nevertheless due, by and large, to Marconi, and we must freely acknowledge that the first success was gained as a result of his ability to shape the whole thing into a practical, usable system, added to the inflexible energy with which he pursued his self-appointed aim.

Marconi's first experiment to transmit a signal by means of Hertzian waves was carried out in 1895. During the fourteen years that have elapsed since then, wireless telegraphy has progressed without pause until it has attained the great importance it possesses today. In 1897 it was still only possible to effect a wireless communication over a distance of 14–20 kilometers. Today, electrical waves are despatched between the Old and the New World, all the larger ocean-going steamers have their own wireless-telegraphy equipment on board, and every navy of significance uses a system of wireless telegraphy. The development of a great invention seldom occurs through one individual man, and many forces have contributed to the remarkable results now achieved. Marconi's original system had its weak points. The electrical oscillations sent out from the transmitting station were relatively weak and consisted of wave trains following each other of which the amplitude rapidly fell—so-called damped oscillations. A result of this was that the waves had a very weak effect at the receiving station, with the further result that waves from various other transmitting stations readily interfered, thus causing disturbance at the receiving station. It is due above all to the inspired work of Prof. Ferdinand Braun that this unsatisfactory state of affairs was overcome. Braun made a modification in the layout of the circuit for the transmission of electrical waves so that it was possible to produce intense waves with very little damping. It was only through this that the so-called long-distance telegraphy became possible, where the oscillations from the transmitting station, as a result of resonance, could exert the maximum possible effect upon the receiving station. The further advantage was obtained that in the main only waves of the frequency used by the transmitting station were effective at the receiving station. It is only through the introduction of these improvements that the magnificent results in the use of wireless telegraphy have been attained in recent times.

Research workers and engineers toil unceasingly on the development of wireless telegraphy. Where this development can lead, we know not. With the results already achieved, however, telegraphy over wires has been extended by this invention in the most fortunate way. Independently of fixed conductor routes and space, we can produce connections between far-distant places, over far-reaching waters and deserts. This is the magnificent practical invention that has issued from one of the most brilliant scientific discoveries of our time!

APPENDIX B

Electrical Oscillations and Wireless Telegraphy—
Ferdinand Braun's Nobel Lecture,
Given at Stockholm on 11 November 1909

In accepting, today, the great honor and privilege of addressing the members of an academy that though of venerable age is constantly renewed and invigorated by the contribution of fresh strength and energy, I hope for your indulgence and understanding when I conceive my task not to be that of talking about wireless telegraphy in general. I have felt it more fitting to limit myself to the narrower field of the activities in which I have been successful in taking some part in the development of the whole.

I shall omit my experiments on the propagation of electrical waves through water, which I carried out in the summer of 1898, and shall turn at once to the experiments that were described and conceived at that time as being transmission through the air.

The following should first be mentioned: Marconi, as far as I know, had begun his experiments on his father's estate in 1895, and continued them in England in 1896. His experiments in La Spezia harbor were, with other ones, carried out in 1897, and a distance of 15 kilometers was attained. In the autumn of the same year, Slaby, using much the same arrangement, reached 21 kilometers over land but only by means of balloons to which were attached wires 300 meters in length. Why, one must ask, was it so difficult to increase the range? If the whole arrangement functioned satisfactorily over a distance of 15 kilometers, why could not double the distance or more be attained by increasing the initial voltage, the means for which were available? It seemed, however, as if ever larger antennae were necessary. It was with this impression—whether the papers had correctly reported the experiments or not, I shall let pass—that I turned my attention to the subject in the autumn in 1898. I set myself the task of obtaining stronger effects from the transmitter.

If I am to give you the general thoughts and concepts that guided me, I must ask you to carry yourselves back with me to the standpoint of our knowledge at that time. What facts were at our disposal and what conclusions could be drawn from them? It was known how sensitive the Hertzian oscillations were to the quality of the spark, and also that lengthening the spark led to definitely deleterious effects whereby the spark became "inactive." In his first paper, Hertz already had called attention to the strong damping of the oscillators and compared their electrical oscillations with the ill-defined acoustic oscillations of wooden rods. Bjerknes, in 1891, had successfully measured the damping and found the logarithmic decrement (a well-known measure for damping) for a linear oscillator to be 0.26, when he used only a minute spark gap. When, however, the spark gap was increased to 5 millimeters, the decrement rose to 0.40. This, and a series of other facts, indicated the existence of strong spark damping. All known facts became understandable if one assumed that at low capacities the spark consumed a great part of the energy, and the longer the spark, the larger was the part of energy it consumed. On the other hand, it had long been known that the discharge of bigger capacities in the customary arcs was always oscillatory and (in radiation-free paths) obviously much less attenuated. In fact Feddersen had photographed directly up to 20 half-cycles of oscillations in 1862. I took hold of this fact.

Considering the greater amounts of energy that can be collected and stored in suitable experimental form in capacitors, one could expect to deliver radiated energy for some time from them. Taken all in all, I concluded that if a *sparkless* antenna could be excited, from a closed Leyden-jar circuit of large capacity, into potential oscillations whose average value was that of the initial charge in a Marconi transmitter, then one would possess a more effective transmitter. There was some doubt whether this could be attained. And further, it was necessary to decide, by experiments on effects at a distance, whether any disturbing factor had been overlooked in these considerations. By suitable dimensioning of the exciter circuit, it was found possible to fulfil the first requirement, and comparative experiments on long-distance effects were in favor of the new arrangement.

Three circuits arose from this, which I described as inductive and direct transmitter excitation, together with a mixed circuit derived from both. In figure I is shown the direct circuit. The transmitter is grounded. In figure II is shown the inductive circuit, in which Marconi's direct grounding is replaced by a "symmetry wire." This name would be entirely suitable if the complete transmitter were floating in free space (for example, in a balloon). The transmitter would then form a half-wavelength, and the excitation point, which should lie at the antinode of the current, would be in the middle. Figure II shows how this circuit is adapted to a mobile station. The set-up is now unsymmetrical due to the proximity of ground. The symmetry wire can be shortened by loading its end with capacity. This arrangement is then known as a counterpoise. It disappears entirely if the connected capacity is infinitely large, that is to say, when the excitation point is on well-conducting ground.

By a suitable design of the Leyden-jar circuit, significantly higher voltages are attained in the transmitter than the charging voltage

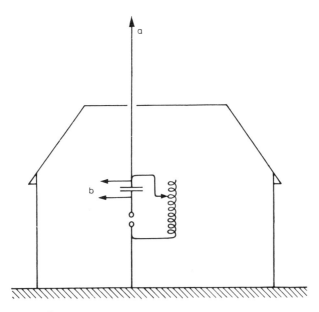

Figure I

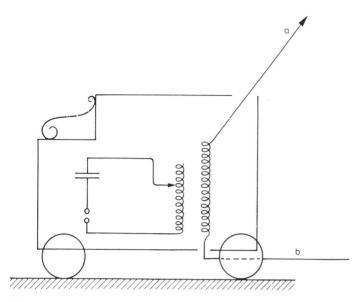

Figure II

of the Leyden-jar circuit. There was some suspicion in my mind that large capacities with bigger spark lengths would behave in the same way as small capacities. At that time, little was known about this. The results of later experiments have, in part, contradicted each other, since other losses appearing with high voltages were overlooked. But as far as the spark resistance was concerned, my fears, as M. Wien recently has shown, were without foundation. Since I wanted to be prepared, however, for every eventuality, I asked myself whether it might not still be possible to increase the power, for instance by connecting *several* circuits of the *same* frequency of oscillation into the excitation circuit of the transmitter. The difficulty was to so couple circuits of this kind together that they would all start to discharge at the same moment, for example within exactly $\frac{1}{10,000,000}$ second. This task occupied me on repeated occasions. One solution, attained in a somewhat different way and to which I was led in the course of my experiments, is given here (figure III). It has been described as an "energy coupling." I will touch later upon the advantages possessed by this arrangement, which remain despite the results obtained by Wien.

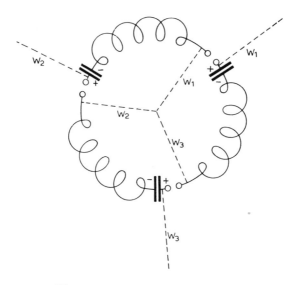

Figure III

The experiments were to be carried further under practical conditions after Easter of 1899. The choice of location for the tests fell to Cuxhaven. In addition to the main task, there was an almost overwhelming pressure of other allied tests and problems; for example, how does the coherer work, particularly under the practical conditions occurring? Is it a resistance or does it behave like a capacitance or both? Can it be replaced by something better defined and, if possible, more quantitatively informative? How do nearby buildings or metal masses such as masts and stays, which play such an important part in practice, affect the antenna? And there was a further multitude of problems with respect to the particular receiving apparatus. And all these problems affected the overall solution so that they all needed to be solved nearly simultaneously. Owing to my professional duties I could devote but little time to the tests, and they were carried on by two of my assistants until the autumn of 1900. The way in which the most favorable conditions were discovered in practice by systematic methods has been described by me elsewhere.

On 16 November 1900, I gave my first public lecture on this subject to the Natural Sciences Society in Strasbourg. There I described, among other matters, the advantages offered by my circuits for tuned telegraphy, advantages that Marconi had by then also recognized. On the following February 1, I demonstrated before the same society the methods on which I had based the tuning of a receiver. I carried out more or less the same experiments before the Assembly of Research Workers in Natural Sciences[*] in Hamburg during the autumn of the same year, as well as demonstrating the practical results on the station at Helgoland.

In the receiver, too, the most important feature was the capacitor circuit, which was directly coupled to the antenna, and which, as I expressed it, collects the energy radiated toward the receiver into the best possible loss-free paths, localizes it, and thus passes it onward in the most suitable form for the detector.

By my arrangements, so-called *coupled* systems were introduced throughout in the wireless-telegraphy system, and at this point we might briefly examine their properties. For preference I have used Oberbeck's pendulum model for illustration, although it does not correspond completely to the electrical conditions. I produce it here (figure IV). Two pendulums of identical frequency are "coupled" through a loaded thread. I draw the first pendulum away from the position of rest and release it. It transmits its energy to the second pendulum, and the latter increases its energy at the cost of the first, exciting pendulum. After some time the whole of the energy appears in the second pendulum. At this point, however, the process repeats itself in the opposite sequence. If I make the first pendulum heavy and the second one light, I can make the oscillation amplitude of the second greater than that of the first. The first pendulum represents the Leyden-jar circuit, the second the transmitter to which—in this case—the whole of the energy of the Leyden-jar circuit is passed. According to the ratio of the capacities the voltage can be amplified (or, if desired, reduced).

In 1895 Oberbeck demonstrated the following by calculation. If a capacitor circuit is allowed to operate inductively upon a second circuit of the *same* natural frequency there appear—most strikingly—*two* oscillations in both circuits, one higher and one

---

[*]The official name of this organization is *Gesellschaft deutscher Naturforscher und Ärzte*, which Braun abbreviates in his speech to *Naturforscherversammlung*. Its correct translation is German Society of Scientists and Physicians.

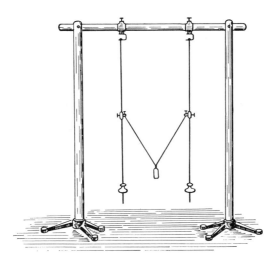

Figure IV

lower than the natural frequency of oscillation. The closer the coupling, that is, the quicker the energy transference from the first to the second circuit becomes, the further apart lie the frequencies. Only for the case of infinitely loose coupling do the two oscillations approximate the natural frequency, that is to say, become equal to each other.

This result holds also for mechanical systems, which include our pendulums. If our two equally tuned pendulums are coupled, then each should exhibit *two* different frequencies of oscillation. The result loses its surprise when the phenomenon is not treated mathematically but, I would like to say, actually takes place before our eyes. The characteristic is this: the oscillations of the second pendulum increase steadily from zero upward, then again decrease, and vice versa. We note from each pendulum what is known in acoustics as "beatings." I shall recall now a method of representing graphically acoustic beating (figure V). As oscillating tuning fork carries a glass plate covered in soot (carbon black). A second tuning fork writes upon this oscillating glass plate by means of a small pin while it itself is drawn across the plate. One tuning fork would describe a curve of constant amplitude (figure VI). The oscillations of both forks are added algebraically. And when both forks have different frequencies, as is here the case, a curve as in figure VII results.

           APPENDIX B

Figure V

Figure VI

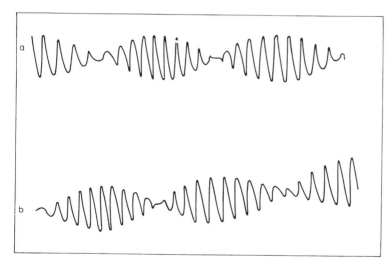

Figure VII

FERDINAND BRAUN'S NOBEL LECTURE

I shall call them briefly "beating (or pulsating) oscillations." Just such curves would arise if we allowed our pendulum to write upon a moving plate. If the exciting pendulum gave the upper curve, *a*, then the excited pendulum would give the lower curve, *b*.

From the elementary law of trigonometry, each such beating oscillation can be considered as arising from the superposition of two harmonic oscillations of different frequency, say n_1 and n_2.

Although this is mathematically possible, experience teaches us that if this pulsating oscillation is applied to a structure capable of oscillation whose natural frequency fits in with one or other of the frequencies n_1 or n_2, then it will be excited in its own natural frequency of oscillation. It selects one of the fictitious harmonic components and endows it thereby with an independent existence. A body so excited is called a resonator, and the phenomenon itself is known as resonance.

The thing is clear to the mind in the case of the tuning-fork example. In space we would observe the beats, but the resonators would separate the two tuning-fork tones. The application to our pendulum model is also obvious. Each part of the system performs pulsating oscillations, so that resonators react to two different harmonic oscillations.* If we wish (and here I return to the electrical example) to record, by means of resonators, oscillations from the radiation sent into space by the antenna, then we have to adjust the resonator to one of the two oscillations.

These electrical oscillations can be separated by means of a variable capacitor circuit, the so-called resonance Leyden-jar circuit (to which I will return later), so long as care is taken that as far as possible the circuit has no feedback into the system being investigated. Oberbeck's result was concerned with the case in which both system components had *closed* current circuits (with quasistationary flow) and were inductively coupled. It can now be easily shown that both oscillations also exist in the open-current path

---

*If time allowed, I would be able to demonstrate both these oscillations "analytically" by means of the pendulum model. This second model, introduced by Mr. Mandelstam, and representing direct coupling through its correct mechanical analogue, allows every detail to be recognized.

APPENDIX B

of an antenna, whether it is excited directly or inductively (I am ignoring higher harmonics).

In the summer of 1902 I was able to erect two experimental stations on two forts at Strasbourg for the purpose of closer study. The task that I had set for us was to determine the most favorable conditions in the receiver. We adopted the resonant circuit, in which known capacitances were combined with calculated self-inductances, so as to bring both parts of the transmitter system into the same natural frequency of oscillation. We fixed likewise the two oscillations arising from the coupling and searched for these with the receiver. The result of the test was, for that time, surprising, as an example will show. If, by means of a coil in the receiver circuit, the oscillations were transferred inductively into a second coil located in a tuned circuit containing the indicator (parallel to a small capacitor), not only the sharpness of the resonance but also—and here was the surprise—the *intensity* of the excitation was raised as soon as the two coils were *moved away* from one another. The intensity increased with increasing distance between the coils, though naturally beyond a certain limit there was again a decrease. Described in the customary expression, the effectiveness increased with *looser* coupling. This result in the receiver was *not* subject to a similar loose coupling in the transmitter.

There were two important results from these experiments: (1) greater freedom from disturbance in the receiver; and (2) a valuable measuring instrument for wireless engineering. When Dr. Franke of Siemens & Halske (who were working with us) saw the tests, he proposed to base an engineering instrument on them. Until then the resonance circuit had been assembled from existing parts to suit the particular requirements, and from whatever came to hand. Through the combination of a Köpsel's calibrated variable rotating capacitor and a number of calculated self-inductances, an apparatus was constructed that covered a large range of wavelengths both conveniently and continuously. The "current effect" was measured by means of a Riess air thermometer, which I had long used for intensity measurements of oscillations. The technical preparation fell to Mr. Dönitz. So arose the wavemeter, described by him and generally named after him, an apparatus that, according to the theory already developed by Bjerknes in 1891, permitted simultaneously the *damping* or attenuation of electrical waves to be measured, a quantity whose numerical value was ever more needed. There are other wavemeters with open current paths; these are simpler, but despite this, because of its other advantages the closed-circuit apparatus has held the field. By means of this instrument the foundations of measuring techniques for wireless telegraphy were laid. It soon displaced our cumbersome laboratory equipment and gave us great help in our scientific investigations, while it became indispensable for rational technical work in the field of electrical oscillations.

In the summer of 1902 came the publication of a theoretical study of the coupled transmitter by Max Wien. This particularly concerned the effect of damping. Wien showed by calculation the versatility of the coupled transmitter. He summed up the qualitative results of his work as follows: "According to the kind of coupling, a powerful but quickly attenuated excitation can be attained that reaches far into the distance, or alternatively a slowly decreasing wavetrain that is capable of exciting a similarly tuned resonator while passing all others by—a cannon shot to be heard afar, or a soft, slowly declining tuning-fork tone."

This theoretical investigation was most effective in clarifying the bases of the problem, and it will remain the foundation. It remains to be seen, however, how closely the data chosen for the numerical examples correspond to actual practice. Some calculated figures and a few laboratory figures were all that were available on the subject of damping. The field of measurements in relation to practice was beginning to be opened up. From then on, the work spread further and further outward, branching into that of the scientific laboratories on the one hand, and the conversion of their results into practice with its complicated conditions and extensive requirements on the other. Success in the latter connection is due to Count Arco and Mr. Rendahl.

The circumstances that led me, more than ten years ago, to introduce the capacitor circuit, have altered greatly in the meantime. The Leyden-jar circuit is still today indispensable in wireless telegraphy. Two properties should be mentioned that I have not yet touched upon:

1. For equal powers it is easier to design an inductor for use with high charging capacities and low voltages than vice versa. This

was a determining factor at the time for the energy circuit mentioned earlier and remained so for this setup.

2. Insulation difficulties are practically nonexistent in the Leyden-jar circuit, but the contrary is the case in the antenna circuit. If, for example, the insulators in a coupled transmitter are damp, the transmitter still works, whereas it can become impossible to charge it statically or with low frequency.

I illustrated the latter point in my lecture in November of 1900 by means of the following experiment. I allowed the transmitter to operate inductively upon a neighboring receiver and so produced current in the latter that brightly lit up an incandescent bulb. I touched the transmitter wire with a moist binding thread that was connected to ground. This had no effect on the operation in the case of the coupled transmitter, but the transmitter with direct inductor charging could not be operated once the damp thread was placed in contact with it.

Before I leave the subject of the coupled system permit me to recall an accessory that was of great use to me and other experimenters. I mean the cathode-ray tube, which I described in 1897. It provided a visual picture of current and voltage waveforms up to 100,000 cycles per second, and was the means by which investigations of period, waveform, intensity, and thereby damping, as well as relative phases could be made.

One of the first applications of this tube was Knut Ångström's neat method of showing directly the hysteresis curve. In a similar way, the permeability of iron up to 130,000 cycles per second was investigated at the Strasbourg institute, and a number of other problems concerned with electrical oscillations were also studied.

Three oscillograms made with the tube will serve to show its application. They illustrate the primary current pattern in the inductor, which interests us, and the significance of the capacitor therein.

In figure VIII the primary current in the noncapacitative circuit falls away relatively slowly when the circuit is broken. On the other hand, if a capacitor (figure IX) is switched in, *oscillations* occur on breaking the circuit. The current falls much more steeply and by nearly twice the value. The secondary coil was open. If

Figure VIII

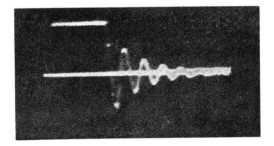

Figure IX

this coil circuit is closed (figure X), the oscillations are faster and are attenuated more strongly.

Many applications of the tube are given in Zenneck's well-known book. I will show you now only the current oscillations in two coupled (but strongly damped) capacitor circuits. You will see that the tubes do in fact show actual beating or pulsating oscillations (figure XI).

In still another place, wireless telegraphy brought me into contact with earlier investigations I had made, this time in connection with work in my youth. I found in 1874 that materials such as galena, pyrite, pyrolusite, tetrahedrite, and so forth, departed from Ohm's law, particularly when an electrode made contact over a small surface area. These materials greatly interested me since they conduct without electrolysis, although they are binary compounds. The resistance appeared to be dependent upon the direction and intensity of the current and I could, for instance, separate the

APPENDIX B

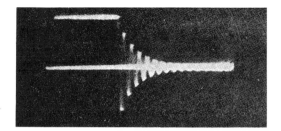

Figure X

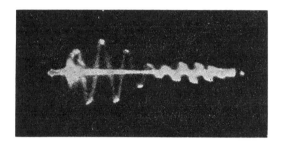

Figure XI

opening and closing currents of a small inductor by means of such materials, in a similar way to that of a Geisler's tube. I did not succeed in finding an "explanation" for the phenomena, for instance, to what material asymmetry the electrical asymmetries (which without doubt existed) corresponded. I had to content myself with showing that the observed phenomena were not brought about through secondary effects such as heating. I was able to demonstrate that it appeared—at least qualitatively—even in ⅟₅₀₀ second, and I was convinced that—perhaps at the furthest limits—an inertialess process was concerned, a view that was supported by E. Cohn while carrying out some other experiments, when he found that the asymmetrical dc resistance could follow current oscillations of 25,000 cycles per second. But always there remained with me a feeling of dissatisfaction, and with it, a faint memory that obviously had never died, but remained half-somnolent at the back of my mind. Instinctively I was driven back to this valve effect (with which I had repeatedly, though in vain, attempted to obtain direct current from oscillations of light) when I began to occupy myself with wireless telegraphy in 1898. The elements showed the expected detector effect, but *at that time* offered no advantages over the coherer. As the swing to aural reception of messages took place, I came back to these materials, and recognized their usefulness for this purpose in 1901. In 1905 the Gesellschaft für drahtlose Telegraphie ("Wireless Telegraphy Co.") decided, on my recommendation, to start up a technical project in this sphere of activity. Today, these detectors—including other combinations of a similar nature—are extensively used. Pierce, by means of the cathode-ray tube, has demonstrated for slow oscillations an almost complete separation of positive and negative current components in the case of molybdenite. It seems to me that it is still an open question whether this will hold also for rapid oscillations.

I shall turn now to another series of experiments.

It had always seemed most desirable to me to transmit the waves, in the main, in one direction only. I shall not concern myself with the successful experiments of this kind made at the Strasbourg Forts in 1901, since it came out later that similar proposals had already been made by others.

I found in 1902 that an antenna, inclined at somewhat less than 10° to the horizon, formed a kind of directional receiver. The receptivity showed a clearly defined maximum for waves passing through the vertical plane in which the antenna was situated. The results were published in March 1903.

A directional transmitter is made up in the following way (figure XII). It is assumed that the antennae A and B, located at corners of an equilateral triangle, are equal in phase, but are delayed by a quarter of a cycle of oscillation relative to antenna C, which is in the third corner. The height CD of the triangle is to be a quarter wavelength. The radiation will then prefer the direction CD. The wave emanating from C will reach AB at the moment that A and B start to oscillate.

The task arose to attain this kind of phase difference for rapid oscillations and, prior to this, to measure such differences. A measuring method came easily to hand, one that has also proved itself in practical experiments. The solution of the other task did not go well using the scheme that I had thought out. On the other hand, two of my assistants found an ingenious solution when they

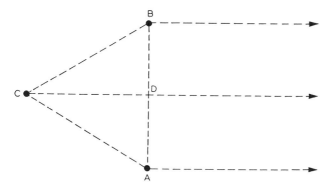

Figure XII

took up the work, at my suggestion, in the Strasbourg institute. Experiments were carried out on a big parade ground in the vicinity of Strasbourg (spring of 1905).

In figure XIII is shown, schematically, the layout used. The field was measured at a fair distance away, that is to say, in the so-called far zone. There was satisfactory agreement between theory and observation, and the results were checked in various ways. Further, it was shown that the experimental layout functioned in the desired sense. By suitable distribution of the amplitudes in the three transmitters, a field as in figure XIV was calculated (the dotted curve is the measured field). The radial vectors represent the range. If the roles of the three transmitters are interchanged—by simply tripping a change-over switch—the preferred direction can be rotated through 120° or 60°.

One is led to the conclusion that the radiation of a transmitter is reduced here by the oscillations in its neighbor, which are shifted in position and phase, a conclusion that could be proved experimentally.

If nowadays optical phenomena are ascribed to electrical molecular resonators, then electrical processes, as demonstrated here by a single example, can also be linked up with optical phenomena, though this can hardly be experimentally verified in this field.

Here, the study of electrical oscillations supplements that of optical oscillations, and since we are in the position to tackle a problem in either field by analogy with a phenomenon that is comprehended in the other field, the first attack on the problem

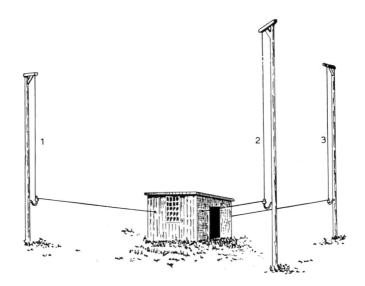

Figure XIII

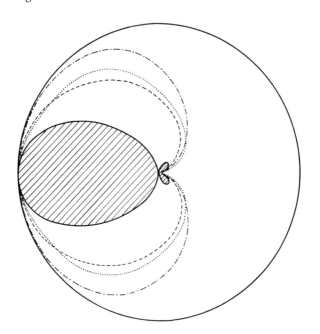

Figure XIV

APPENDIX B

can be made from the electrical or the optical standpoint according to whichever presents the easier concept to realize. Perhaps I can illustrate this by means of two worked-out examples.

Elementary considerations led me to the conclusion that a medium, composed of layers of different dielectric constants, must behave as a uniaxial crystal if it is assumed that the layer thicknesses are only a fraction of a wavelength. I was able to confirm this conclusion in the following way (figure XV). A beam of practically parallel electrical rays emerges from the Hertzian reflector. It strikes a structure made of bricks in layers having the same breadth of air layers. These layers lie open to short waves, but if the wavelength is about twelve times longer than the layer thickness, then the brick grating behaves toward it as a body that homogeneously occupies the space but exhibits double refraction. The electrical oscillations are linearly polarized and incident at an azimuth of $45°$ upon the brick layers. A brick structure that is about 2.5 times the thickness of a single brick has the effect of a quarter-wave foil of mica, and the linearly incident ray emerges circularly polarized, as we deduce from the investigations with a Righi resonator. Assume it is right-handed circular. If the layer thickness is now doubled, the emergent wave is again linearly polarized, though in the other quadrant. And so we can transform the ray, by continuous addition of further thicknesses, into a left-handed circular one, and finally back into a ray that is linearly polarized parallel

Figure XV

to the incident ray. The double refraction of the brick grating surpasses that of calcite. Optically, this brick structure would correspond to a tiny crystal of a few thousandths of a millimeter length of edge, but electrically it is 2.5 meters thick, weighs 4,000 kilograms, and its raw material is worth about 200 marks. The analogue of a corresponding optical phenomenon was also demonstrated by me at a later date.

This phenomenon of double refraction does not depend upon the use of rigid materials. Whether the double refraction occurring in cross-striated muscle results from a similar layer structure is thus a closely related question.

We have studied so far an electrically unknown but optically conjectured phenomenon, and both have been discovered to exist. The following example is concerned with demonstrating the unknown optical phenomenon corresponding to a known electrical phenomenon. It seemed to me to be of interest to reproduce the Hertzian grid experiment in the field of visible rays. For this to be realized, a very fine grating of metal wires was necessary, and from 10,000 to 100,000 tiny wires, separated by air gaps, had to be located within a width of 1 millimeter. Mechanical methods of manufacture are impossible, but a Hertzian grid could be made in the following way. If a powerful discharge is passed through a thin metal wire on a glass plate, or between two such plates, the well-known sputtering or vaporization effect occurs, as you can see from figure XVI. The metal wire vaporizes (temperatures of up to 30,000° are calculated). The metal vapor is driven outward by the pressure arising from the explosive effect (figure XVII) and is then again precipitated obviously in a kind of grid structure on

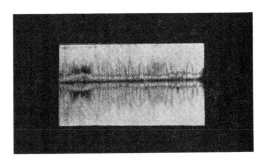

Figure XVI

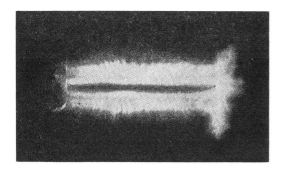

Figure XVII

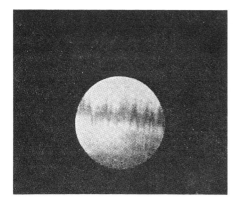

Figure XVIII

the glass. If we allow linearly polarized light to fall upon the prepared surface, it will, if the oscillations are parallel to the lines of the grid, be strongly reflected and strongly absorbed—the preparation appears dark (figure XVIII). If the plane of the oscillations is turned so that it is perpendicular to the lines of the grid, the metal layer becomes transparent (figure XIX). We have the complete optical analogue to a Hertzian grid made out of moderately good conductors.

This experiment permits a further development. If we imagine that in an organized fabric such as muscle tissue, plant fibers, and so forth, there exists a similar fine grating structure somewhat in the form of the finest possible channels, then if we could succeed in filling these with metal, the preparation would have the optical effect of a Hertzian grid. In 1896 H. Ambronn, treating the above-

Figure XIX

mentioned substances with gold or silver salts, discovered phenomena that I explained in this way. In an exhaustive investigation into this matter I have found confirmation everywhere of my concept, and nowhere a contradiction. Yet a direct and incontrovertible proof would be extremely desirable because of the importance of its consequences. For if my idea is, as I believe, correct, then we would in this way not only discover submicroscopic gratings, but, as a result of electrical imitation, we would even be able, to some extent, to make a picture of the material structure, which is as yet invisible to the human eye. This method would augment those so far available in a most valuable manner, for it takes its place just where the microscope and—because of the density of the particles—even the ultramicroscope, reach the limit of their capacity.

I must now finish this address. The sputtering experiments led me back to the Leyden-jar circuit. I pursued for a long time the aim of automatically switching out the Leyden-jar circuit from the oscillating system as soon as it had given up its energy to the secondary conductor. I attempted this in the following way. A thin wire was connected into the Leyden-jar circuit, and I hoped that, at the right moment, the primary circuit would be switched out as a result of vaporization of the wire. The experiment was not successful, at any rate at the frequencies I used, apparently because the highly heated metal vapor remained ionized for too long a time. The problem was solved, however, by Max Wien using the so-called quenched spark, and by Rendahl using the mercury spark gap. Practical experience has augmented Wien's discovery. Arising from this and through the agency of Rendahl and Arco came the so-called tone spark. The small hissing or quenched sparks of Wien meet the conditions that I had hoped to produce artificially. The Leyden-jar circuit cuts itself out at the most suitable moment, and the greater part of the primary energy then oscillates in the highly conductive paths in the transmitter at its own natural frequency.

On the occasion of my first lecture in November 1900 I closed with the following words: "Sometimes, wireless telegraphy has been described as spark telegraphy, and so far a spark in one place or another has been unavoidable. Here, however, it has been made as harmless as possible. This is important. For the spark that produces the waves also destroys them again as Saturn destroyed his own children. What was pursued here could be truthfully described as *sparkless telegraphy*."

Finishing as I did with these words at that time, I feel happy to think that with the means I have described we have come appreciably nearer to this target, and have thereby made the coupled transmitter still more effective.

CHRONOLOGY

1850
Born in Fulda, 6 June

1856–1868
At school in Fulda

1868–1870
Student in Marburg and Berlin

1870-1874
Assistant in Berlin and Würzburg

1872
Awarded doctorate under Helmholtz in Berlin, 23 March

1870–1874
Investigations of the vibrations of strings and of electrolytic conduction

1872–1874
Literary activity on the *Fliegende Blätter*

1874–1877
Teacher at Thomas *Gymnasium* in Leipzig

1874
Publication of the discovery of the rectifier effect in semiconductors, 23 November

1875
Book, *Der junge Mathematiker und Naturforscher*

1877–1879
Associate professor in Marburg

1878
Refutation of Thomson-Helmholtz theory for calculation of electromotive force

1880–1882
Associate professor in Strasbourg

1883
Conclusion of rectifier-effect investigations

1883–1885
Professor in Karlsruhe

1883–1885
High-temperature physics; investigations of thermoelectricity

1884
Electrical pyrometer

1885
Marriage to Amalie Bühler of Lahr in the Black Forest, 23 May

1885–1895
Professor in Tübingen

1885–1889
Planning and construction of the Physics Institute in Tübingen

1885–1887
Investigations of solubility and compressibility

1887
Le Châtelier-Braun principle; Braun electrometer

1888
Discussion with Heinrich Hertz; "deformation currents"

1890
Bore hole near Sulz on the Neckar

1895–1918
Professor in Strasbourg

1897
Description of the cathode-ray oscilloscope, 15 February

1897
First trip to America; demonstration of the cathode-ray oscilloscope in Toronto

1897
Wireless telegraphy through the earth and through water, November

1898
First demonstration of the Braun transmitter in Strasbourg, 20 September

1898
Braun transmitter patent applied for, 14 October

1898
Call to Leipzig declined

1898
Funkentelegraphie GmbH founded in Cologne, 15 December

1899–1900
Radio experiments near Cuxhaven

1899
Telebraun founded in Hamburg, 7 July

1899
Crystal detector, multiple tuning circuits, feedback circuit

1900
Trip to the Sahara

1900
Contact established between Helgoland and Cuxhaven, 24 September

1900
First public lecture on wireless telgraphy, 16 November

1901
Braun-Siemens Gesellschaft founded in Berlin, July

1901–1906
Experiments with directional transmission on Strasbourg drill ground

1901–1914
Theoretical development of high-frequency physics in Strasbourg (the Braun school)

1902
Scientists take Braun's part in the quarrel with Slaby in Karlsbad

1903
Kaiser Wilhelm II asks German wireless-telegraphy experts to unite, April

1903
Telefunken founded in Berlin, 27 May

1903–1905
Proof of essential equivalence of light and electromagnetic waves; grating experiments

1905
Lecture before Wilhelm II in Berlin, 16 February

1905
Declined call to Berlin

1905–1906
Rektor of the University of Strasbourg

1906
Operation in Freiburg, January

1905–1908
Albert Schweitzer Braun's student

1909
Award of Nobel Prize in Stockholm, 10 December

1911
Construction of wireless station at Sayville, Long Island

1913
Investigation of frame antenna completed

1913
Grand tour of Italy, September

1914
Experiments with Graf Zeppelin on Lake Constance, April

1914
World War I declared

1914
Second trip to America; American Marconi Co. lawsuit against Atlantic
Communication Co.

1915
Honored by Institute of Radio Engineers in New York, January

1916–1917
Physics for Women

1917
Honorary doctorate, University of Vienna

1917
Death of Amalie Braun, 17 September

1918
Died in Brooklyn, 20 April

BIBLIOGRAPHY

A. FERDINAND BRAUN'S WRITINGS

1865–1867

1. Das Wasser [Water], *Kurhessische Schulzeitung* (Kassel).

1a. Eine etwas verkürzte Darstellungsweise des Rhodanammoniums und einiger anderer Rhodanmetalle [A somewhat shortened method of producing ammonium thiocyanate and some other metallic thiocyanates], *Chem. Central-Blatt* 11:245–246 (1866); *Pharm. Zeitschrift Russland* 5:331–332 (1866).

1b. Die Trinkbarmachung des Seewassers auf chemischem Wege [Making sea water potable by chemical means], *Chem. Central-Blatt* 12:241–251 (1867).

1872

2. Ueber den Einfluss von Steifigkeit, Befestigung und Amplitude auf die Schwingungen von Saiten [On the effect of stiffness, fastening, and amplitude on the vibration of strings], doctoral diss., University of Berlin; *Ann. Phys. Chem.* (2)147:64–91.

1874

2a. Ueber elastische Schwingungen, deren Amplituden nicht unendlich klein sind [On elastic vibrations of noninfinitesimal amplitudes], *Ann. Phys. Chem.* (2)151:51–69.

3. Ueber die galvanische Leitungsfähigkeit geschmolzener Salze [On the galvanic conductivity of melted salts], *Berichte deutsch. chem. Ges.* 7:958–962; *Ann. Phys. Chem.* (2)154:161–196 (1875).

4. Ueber die Stromleitung durch Schwefelmetalle [On the conduction of electrical currents through metallic sulfides], *Ann. Phys. Chem.* (2)153:556–563.

1875

5. Ueber die unipolare Electricitätsleitung durch Gasschichten von verschiedener Leitungsfähigkeit [On unipolar electric conduction through gas layers of various conductivities], *Ann. Phys. Chem.* (2)154:481–507.

6. *Der junge Mathematiker und Naturforscher* [The Young Mathematician and Scientist], Leipzig: Spamer, 420 pp.

1876

7. Ueber die Natur der elastischen Nachwirkung [On the nature of elastic recovery], *Sitzungsber. Naturf. Ges. Leipzig* 3:28–35.

8. Versuche über Abweichungen von Ohm'schen Gesetz in metallisch leitenden Körpern [Experiments on deviations from Ohm's law in metallically conducting solids], *ibid.* 3:49–62.

1877

9. Abweichungen vom Ohm'schen Gesetz in metallisch leitenden Körpern [Deviations from Ohm's law in metallically conducting solids], *Ann. Phys. Chem.* (3)1:95–110.

1878

10. Bemerkungen über die unipolare Leitung der Flamme [Remarks on the unipolar conduction of the flame], *ibid.* (3)3:436–447.

11. Ueber unipolare Electricitätsleitung [On unipolar electric conduction], *ibid.* (3)4:476–484; *Sitzungsber. Ges. Naturwiss. Marburg.*

12. Ueber die Electricitätsentwickelung als Aequivalent chemischer Processe [On the generation of electricity as an equivalent of chemical processes], *Ann. Phys. Chem.* (3)5:182–215.

1879

13. Ueber Kugelfunktionen [On spherical functions], *Sitzungsber. Ges. Naturwiss. Marburg.*

14. Ueber elliptische Schwingungen [On elliptic oscillations], *ibid.*

1882

15. Electrochemische Untersuchungen [Electrochemical investigations], *ibid.*

16. Ueber galvanische Elemente, welche angeblich nur aus Grundstoffen bestehen, und den Nutzeffekt chemischer Processes [On galvanic elements supposed to consist entirely of pure chemical elements, and the efficiency of chemical processes], *Ann. Phys. Chem.* (3)17:593–642.

1883

17. Einige Bemerkungen über die unipolare Leitung fester Körper [Some remarks on unipolar conduction in solids], *Ann. Phys. Chem.* (3)19:340–352.

18. Ueber Ziel und Methode des electrotechnischen Unterrichts [On the goal and method of instruction in electrical engineering], *Karlsruher Zeitung*.

19. Die Internationale elektrische Ausstellung Wien 1883 [International Electrical Exposition, Vienna, 1883], *Centralzeitung für Optik und Mechanik*.

1885

20. Über die Thermoelectricität geschmolzener Metalle [On the thermoelectricity of molten metals], *Sitzungsber. Akad. Wiss. Berlin* 289–298.

1886

21. *Ueber Gesetz, Theorie und Hypothese in der Physik* [On Law, Theory, and Hypothesis in Physics], Tübingen: Franz Fues, 23 pp.

1887

22. Untersuchungen über die Löslichkeit fester Körper und die den Vorgang der Lösung begleitenden Volum- und Energieänderungen [Investigations of the solubility of solids and the related changes in volume and energy], *Sitzungsber. Akad. Wiss. München* 16:192–219; *Ann. Phys. Chem.* (3)30:250–274; *Zeitschrift physikal. Chem.* 1:259–269, 269–272.

23. Ueber die Abnahme der Compressibilität von Chlorammoniumlösung mit steigender Temperatur [On the decrease of compressibility of a solution of ammonium chloride with increasing temperature], *Ann. Phys. Chem.* (3)31:331–335.

24. Bemerkung über den Zusammenhang der Compressibilität einer Lösung mit derjenigen der Bestandtheile [Remark on the relationship of a solution's compressibility with that of its constituents], *ibid.* (3)32:504–508.

25. Ueber einen allgemeinen qualitativen Satz für Zustandsänderungen nebst einigen sich anschliessenden Bemerkungen, insbesondere über nicht eindeutige Systeme [On a general qualitative law for state changes, together with a few related remarks, especially about ill-defined systems], *Nachrichten Ges. Wiss. Göttingen* 448–462; *Ann. Phys. Chem.* (3)33:337–353 (1888).

26. Ueber das electrische Verhalten des Steinsalzes [On the electric behavior of rock salt], *Ann. Phys. Chem.* (3)31:855–872; (3)32:700.

27. Ein Versuch über Lichtemission glühender Körper [An experiment on the light emission of glowing solids], *Nachrichten Ges. Wiss. Göttingen* 465–467; *Ann. Phys. Chem.* (3)33:413–415 (1888).

28. Bemerkung über die Erklärung des Diamagnetismus [Remark on the explanation of diamagnetism], *ibid.* 462–465; (3)33:318–322 (1888).

29. Ueber die Volumänderung von Gasen beim Mischen; ein Beitrag zur Frage, ob der Druck eines gesättigten Dampfes im Vacuum ein anderer ist, als in einem Gase [On changes of volume in gases as they are mixed; contribution to the question whether the pressure of a saturated vapor in vacuum differs from that of a gas]; *Ann. Phys. Chem.* (3)34:943–952.

1888

30. Über ein elektrisches Pyrometer für wissenschaftliche und technische Zwecke [On an electric pyrometer for scientific and technological purposes], *Elektrotech. Zeitschrift* 9:421–426.

31. Berichtigung, die Compressibilität des Steinsalzes betreffend [Correction concerning the compressibility of rock salt], *Ann. Phys. Chem.* (3)33:239–240.

32. Über elektrische Ströme, enstanden durch elastische Deformation [On electric currents generated by elastic deformation], *Sitzungsber. Akad. Wiss. Berlin* 895–903; *Ann. Phys. Chem.* (3)37:97–107 (1889).

33. Über Deformationsströme; insbesondere die Frage, ob dieselben aus magnetischen Eigenschaften erklärbar sind [On deformation currents; especially the question whether they can be explained from magnetic properties], *ibid.* 959–975; (3)37:107–127 (1889).

1889

34. Nachtrag zu meinem Aufsatz 'Untersuchungen über die Löslichkeit etc.' [Addendum to Ref. 22], *Ann. Phys. Chem.* (3)36:591.

35. Über Deformationströme: dritte Mittheilung [On deformation currents: third communication], *Sitzungsber. Akad. Wiss. Berlin* 507–518; *Ann. Phys. Chem.* (3)38:53–68.

1890

36. Bemerkung über Deformationsströme [Remark on deformation currents], *Ann. Phys. Chem.* (3)39:159–160.

37. Ein Comparator für physikalische Zwecke [A comparator for physical purposes], *ibid.* (3)41:672–630.

38. Ueber Tropfelectroden [On drop electrodes]; *ibid.* (3)41:448–462.

39. Beobachtungen über Electrolyse [Observations on electrolysis], *Sitzungsber. Akad Wiss. Berlin* 1211–1222; *Ann. Phys. Chem.* (3)42:450–464 (1891).

1891

40. Kapillarität [Capillarity], in A. A. Winkelmann, Ed., *Handbuch der Physik,* Leipzig.

41. Ueber einfache absolute Electrometer für Vorlesungszwecke [On simple absolute electrometers for lecture purposes], *Zeitschrift für den physikalischen und chemischen Unterricht;* Ueber absolute Vorlesungselectrometer, *Ann. Phys. Chem.* (3)44:771–773; Absolute electrometer for lecture purposes, *Nature* 46:150.

42. Über die Verwandlung chemischer Energie in elektrische [On the transformation of chemical energy into electrical], *Elektrotech. Zeitschrift* 12:673–674.

43. Zur Berechnung der electromotorischen Kraft inconstanter Ketten [On the computation of the electromotive force of inconstant chains], *Ann. Phys. Chem.* (3)44:510–512.

44. Ueber Electrostenolyse [On electrostenolysis], *ibid.* (3)44:473–500.

45. Ueber electrocapillare Reactionen [On electrocapillary reactions], *ibid.* (3)44:501–509.

1892

46. (With Karl Waitz:) Beobachtungen über die Zunahme der Erdtemperatur, angestellt im Bohrloch zu Sulz am Neckar [Observations on the rise in the temperature of the earth made in the drill hole at Sulz on the Neckar], *Jahreshefte des Vereins für vaterländische Naturkunde in Württemberg* 48:1–12.

47. Bemerkung zu der Erwiderung des Herrn Pellat, das Gesetz über die Gleichheit der Potentiale beim Uebergang von einem Metalle zu der Lösung eines seiner Salze betreffend [Remark on M. Pellat's reply concerning the law of equal potentials in the transformation of a metal into a solution of one of its salts], *Ann. Phys. Chem.* (3)45:185–186.

48. *Ueber elektrische Kraftübertragung, insbesondere über Drehstrom* [On the Transmission of Electrical Energy, Especially by Alternating Current], Tübingen: H. Laupp, 38 pp.; Ueber die Lauffener elektrische Kraftübertragung [On the transmission of electrical energy from Lauffen], *Jahreshefte des Vereins für vaterländische Naturkunde in Württemberg* 48:lxviii–lxix.

1893

49. *Das Physikalische Institut der Universität Tübingen* [Physics Institute at the University of Tübingen], Tübingen: University memorial brochure.

50. Zur physikalischen Deutung der Thermoelectricität [Toward a physical interpretation of thermoelectricity], *Ann. Phys. Chem.* (3)50:111–117.

1894

51. Ueber die kontinuierliche Elektrizitätsleitung durch Gase [On continuous electrical conduction through gases], *Zeitschrift physikal. Chem.* 13:155–162.

1895

52. Thermoelektrizität [Thermoelectricity], in A. A. Winkelmann, Ed., *Handbuch der Physik,* Leipzig.

1896

53. Versuche zum Nachweis einer orientirten electrischen Oberflächenleitung [Experiments toward a proof of a directional electrical surface conduction], *Nachrichten Ges. Wiss. Göttingen* 157–165; *Ann. Phys. Chem.* (3)59:673–681.

54. Ueber den continuirlichen Uebergang einer electrischen Eigenschaft in der Grenzschicht von festen und flüssigen Körpern [On the continuity of an electric property at the boundary layer between solids and liquids], *ibid.* 166–171; (3)59:682–687.

55. Ueber die Leitung electrisirter Luft [On the conduction of electrified air], *ibid.* 172–176; (3)59:688–692.

56. Ein Versuch über magnetischen Strom [An experiment on magnetic current], *ibid.* 177–178; (3)59:693–694.

56a. Ueber die Natur des Flüssigkeitszustandes [On the nature of the liquid state], *Deutsch. naturf. Verhandlungen* Pt. 2 (First Half), 62.

1897

57. Ueber ein Verfahren zur Demonstration und zum Studium des zeitlichen Verlaufs variabler Ströme [On a method of demonstrating and studying the time dependence of variable currents], *Ann. Phys. Chem.* (3)60:552–559.

58. Ueber Bewegungen, hervorgebracht durch den electrischen Strom [On motions caused by electric current], *Ann. Phys. Chem.* (3)63:324–328.

58a. Demonstrations on the form of alternating currents [in German], *Brit. Assn. Rep.* 67:570.

58b. (With J. E. Myers:) On the decomposition of silver salts by pressure, *Proc. Phys. Soc.* 15:200–202; *Phil. Mag.* (5)44:172–173.

1898

59. Erwiderung [Reply, to A. Hess's 'Reclamation' (cf. Refs. 269–270 below], *Ann. Phys. Chem.* (3)65:372–373.

60. Notiz über Thermophonie [Note on thermophony], *ibid.* (3)65:358–360.

61. Ueber Lichtemission an einigen Electroden in Electrolyten [On the luminosity of electrodes immersed in electrolytes], *ibid.* (3)65:361–364.

62. Ein Kriterium, ob eine leitende Oberflächenschicht zusammenhängend ist und über die Dampfspannung solcher Schichten [A criterion for the cohesion of a superficial conducting film and on the vapor pressure of such films], *ibid.* (3)65:365–367.

63. Zeigen Kathodenstrahlen unipolare Rotation? [Do cathode rays exhibit unipolar rotation?], *ibid.* (3)65:368–371.

64. *Telegraphier-System ohne fortlaufende Leitung* [System of Telegraphy Without Continuous Transmission Line], German Patent 115,081, 12 July.

65. *Schaltungsweise des mit einer Luftleitung verbundenen Gebers für Funkentelegraphie* [Method of Connecting a Wireless-telegraphy Dispenser Coupled to an Aerial Transmission], German Patent 111,578, 14 October.

66. Über die Entstehung rotierender Magnetfelder durch Foucaultströme und über Methoden zur übersichtlichen Prüfung von Wechsel- und Drehfeldern [On the generation of rotating magnetic fields by Foucault currents and on methods for examining alternating and rotational currents at a glance], *Elektrotech. Zeitschrift* 19:204–206.

1898

67. *Über physikalische Forschungsart* [Methods of Research in Physics], Strasbourg: Heitz, 31 pp.

1901

68. *Drahtlose Telegraphie durch Wasser und Luft* [Wireless Telegraphy Through Water and Air], Leipzig: Veit, 68 pp.

69. Über drahtlose Telegraphie [On wireless telegraphy], *Naturforscherversammlung* (Hamburg), September 24; *Physikal. Zeitschrift* 3:143–148 (1902).

69a. Wireless telegraphy, *Electrician* 46:778–789.

70. Über rationelle Senderanordnungen für drahtlose Telegraphie [On rational transmitter circuits for wireless telegraphy]. *Physikal. Zeitschrift* 2:373–374.

71. Über einige Sendervarianten für drahtlose Telegraphie [On some alternate connections for wireless telegraphy], *Elektrotech. Zeitschrift* 22:469–470.

1902

72. Drahtlose Telegraphie [Wireless telegraphy], *Umschau* (Frankfurt), 22 pp.

73. Über die Erregung stehender elektrischer Drahtwellen durch Entladung [On the generation of stationary electric waves on wires through condenser discharge], *Ann. Phys.* (4)8:199–211.

74. Entgegnung auf die Bemerkung von Herrn Slaby [Rebuttal of Herr Slaby's remark], *ibid.* (4)9:1334–1338.

1903

75. Entwicklung und heutiger Stand der drahtlosen Telegraphie [Development and present state of wireless telegraphy], *Illustrirte Zeitung* (Berlin).

76. Erklärung auf Herrn Slabys Antwort [Clarification regarding Herr Slaby's reply], *Ann. Phys.* (4)10:665–672.

77. Einige Versuche über Magnetisierung durch schnelle Schwingungen [Some experiments on magnetization by high-frequency oscillations], *ibid.* (4)10:326–333.

78. Notizen über drahtlose Telegraphie [Notes on wireless telegraphy], *Physikal. Zeitschrift* 4:361–364.

79. Article in *Die Woche*.

1904

80. Hermann Georg Quincke, *Ann. Phys.* (4)15(No. 13):i–viii.

81. Methoden zur Vergrösserung der Sendeenergie für drahtlose Telegraphie [Methods for increasing transmitter energy of wireless telegraph systems], *Physikal. Zeitschrift* 5:193–199; *Electrician* 53:19–21 (22 April).

82. Sender mit verminderter Dämpfung [Transmitter with decreased damping], *Elektrotech. Zeitschrift* 25:1121–1122.

83. Herstellung doppelt brechender Körper aus isotropen Bestandtheilen [Birefringent substances formed from isotropic constituents], *Physikal. Zeitschrift* 5:199–203.

84. Drahtlose Telegraphie [Wireless telegraphy], *Elektrotech. Zeitschrift*.

1905

85. Neuere Methoden und Ziele der drahtlosen Telegraphie [New methods and goals of wireless telegraphy], *Jahrbuch der Schiffbautechnischen Gesellschaft* 6:107–134.

86. *Über drahtlose Telegraphie und neuere physikalische Forschungen* [On Wireless Telegraphy and Recent Physical Research], Strasbourg, 32 pp.

87. Der Hertzsche Gitterversuch im Gebiet der sichtbaren Strahlung [Hertz's grating experiment with visible radiation], *Sitzungsber. Akad. Wiss. Berlin* 4:154–167 (1904); *Ann. Phys.* (4)16:1–19.

88. Über metallische Gitterpolarisation, insbesondere ihre Anwendung zur Deutung mikroskopischer Präparate [Polarization by metallic gratings, especially its application in the interpretation of microscopic specimens], *Ann. Phys.* (4)16:238–277.

89. Einige Beobachtungen, die sich auf künstliche Doppelbrechung beziehen [Some observations regarding artificial birefringence], *ibid.* (4)16:278–281.

90. Optische Doppelbrechung in isotropen, geschichteten Medien [Optical birefringence in isotropic layered media], *ibid.* (4)17:364–366.

91. Mechanismus der galvanischen Zerstäubung [Mechanism of electric pulverization], *ibid.* (4)17:359–363.

92. Zur Aufklärung eines Missverständnisses betreffend Dämpfung elektrischer Wellen [On the clarification of a misunderstanding relating to electric waves], *Elektrotech. Zeitschrift* 26:87.

93. Einrichtung, um im Vakuum Entfernungen ändern zu können [Device for adjusting distance in vacuum], *Ann. Phys.* (4)16:416.

1906

94. On directed wireless telegraphy, *Electrician* 57:222–224, 244–248.

94a. High-frequency oscillations, *ibid.* 56:546–549; *Écl. Él.* 46:311–315.

1907

95. Gerichtete drahtlose Telegraphie [Directed wireless telegraphy], *Jahrbuch der drahtlosen Telegraphie und Telephonie* 1.

95a. Radio-telegraphic convention, *Electrician* 59:565–566.

1910

96. Elektrische Schwingungen und drahtlose Telegraphie [Electrical oscillations and wireless telegraphy], in *Les prix Nobel en 1909,* Stockholm: Norstedt & Soner, 18 pp., *Jahrbuch der drahtlosen Telegraphie und Telephonie* 4:1–20.

97. Über das sogenannte Le Châtelier-Braunsche Prinzip [On the so-called Le Châtelier-Braun principle], *Ann. Phys.* (4)32:1102–1106.

98. Erzeugung polarisierten Lichtes an sehr feinen dielektrischen Gittern [Diffraction of polarized light by very fine dielectric gratings], *Physikal. Zeitschrift* 11:817–822.

1913

99. Über den Ersatz von offenen Strombahnen in der drahtlosen Telegraphie durch geschlossene [On replacing open wireless-telegraphy circuits by closed ones], *Jahrbuch der drahtlosen Telegraphie und Telephonie* 7.

100. Eine absolute Messung des vom Eiffelturm ausstrahlenden Feldes in Strassburg [An absolute measurement of the field radiated from the Eiffel Tower made in Strasbourg], *ibid.*

1914

100a. Letter to the editor, *ibid.* 8:475–477.

1915

101. Die drahtlose Telegraphie [Wireless telegraphy], in E. G. Warburg, Ed., *Physik* (Kultur der Gegenwart series), Leipzig: Teubner.

B. UNPUBLISHED WRITINGS, LETTERS, DIARIES OF FERDINAND BRAUN

[DM signifies items in possession of Deutsches Museum in Munich.]

1865

102. *Krystallographie* [Crystallography], manuscript lost.

1868–1878

103. Letters to parents (in possession of Mrs. Ruth Stadler in Pforzheim).

104. Letters (DM and autograph collection Darmstädter in Deutsche Staatsbibliothek in Berlin).

1913–1918

105. Diaries and letters (in possession of Mrs. Konrad F. Braun in Kingston, RI).

1917–1918

106. Über die absolute Grösse der Molekularkräfte und die Natur des Flüssigkeitzustandes [On the absolute magnitude of molecular forces and the nature of the liquid state](DM).

107. The law of the state of fluidity (DM).

108. Über die Natur der photochemischen, insbesondere der photographischen Prozesse [On the nature of photochemical and especially photographic processes](DM).

109. Dichroitische Metallhäute [Dichrotic metallic films](DM).

110. Untersuchungen über dünne Silberschichten durch Reduktion von dünnen Halogen-Silberspiegeln gewonnen [Investigations of thin silver films obtained by reduction of thin halogen silver mirrors](DM).

111. On the law of attraction of molecules (DM).

112. On electro-stenolysis; an attempt to explain the phenomenon (DM).

113. *Physics for Women* (DM).

C. WRITINGS ABOUT FERDINAND BRAUN

114. Georg Graf Arco, Prof. Dr. Ferdinand Braun † [Obituary], *Jahrbuch der drahtlosen Telegraphie und Telephonie* 13 (1918).

115. Philipp Braun, Grabrede [Funeral speech], *Fuldaer Tageblatt,* 6 June 1921.

116. J. S. Dieckmann, Ferdinand Braun, *Der blaue Punkt* (Hildesheim), January 1958.

117. A. A. Eichenwald, Ferdinand Braun, *Elektrichestvo* (1932).

118. Adolf Franke, Ferdinand Braun, *Elektrotech. Zeitschrift* 39:269 (1918).

119. Franz Graser, Der Fuldaer Nobelpreisträger Ferdinand Braun und seine Ahnen [Fulda's Nobel Prize-winner Ferdinand Braun and his ancestors], *Hessische Familienkunde* (Frankfurt), 1957 and 1969.

120. P. Lazarev, Ferdinand Braun, *Uspekhi fyzicheskikh nauk* 1 (1918).

121. George Lewis and F. J. Mann, Ferdinand Braun—Inventor of the cathode-ray tube, *Electrical Communication* 25:319–327 (1948).

122. Erich Maendl, Karl Ferdinand Braun, *Via Regia* (Munich and Vienna, 1955.

123. L. Mandelstam and N. Papalexi, Ferdinand Braun zum Gedächtnis [In memoriam Ferdinand Braun], *Naturwiss.* 16:621–626 (1928).

124. Heinrich Meyer, Wem verdanken wir Rundfunk und Fernsehen? [Who brought us radio and television?], *Werkzeitschrift der Fortuna-Werke* (Stuttgart), 1958.

125. Ernst Nebhut, Braun (broadcast script, Berlin), 1943.

126. Carl Piel, Ferdinand Braun, der letzt Strassburger Physiker [Ferdinand Braun, last Strasbourg physicist], *Kölner Volkszeitung,* 1940.

127. E. W. Robbins, Electronic magic and its unknown inventor, *Sylvania Beam* (New York), 1949.

128. Hermann Rohmann, Ferdinand Braun † [Obituary], *Physikal. Zeitschrift* 19:537–539 (1918).

129. W. F. Schrage, Wizard hated mathematics, New York *Sun,* 20 May 1939.

130. Joseph Strieb, Erinnerungen an Ferdinand Braun [Reminiscences of Ferdinand Braun], *Strassburger Post,* 26 June 1918.

131. Jonathan Zenneck, Ferdinand Braun, *Der deutsche Rundfunk* (Berlin), 1924.

132. Jonathan Zenneck, Ferdinand Braun, in *Lebensbilder aus Kurhessen und Waldeck* (Marburg), 1940.

133. Jonathan Zenneck, Zum 50jährigen Jubiläum der Braunschen Röhre [On the 50th anniversary of the cathode-ray oscilloscope], *Physikalische Gesellschaft* (Berlin), 1947.

134. Jonathan Zenneck, Karl Ferdinand Braun, in *Neue Deutsche Biographie*, Berlin: Duncker & Humblot, 1955; vol. 2, pp. 554–555.

135. Jonathan Zenneck, *Erinnerungen eines Physikers* [Recollections of a physicist], typescript, Munich, 1961.

D. CONTEMPORARY AND UNIVERSITY HISTORY

143. *Allgemeine deutsche Biographie,* Leipzig: Duncker & Humblot, 1875–1900.

144. *Deutsches biographisches Jahrbuch,* Berlin: Reimer, 1897–1932.

145. Address books of the city of Fulda.

146. Book section and historical supplement of *Fuldaer Zeitung.*

147. *Fuldaer Kreisblatt,* several years.

148. Arno Hartmann, *Zeitgeschichte von Fulda,* Fulda, 1895.

149. Vergangenheit spricht zur Gegenwart, historical supplement of *Fuldaer Volkszeitung.*

150. *Hessenland,* several years.

151. Die Ereignisse in Kurhessen, *Leipziger illustrierte Zeitung,* 1850.

152. Philipp Lotz, Die Strafbayern, *Frankfurter Zeitung,* 1911.

153. Aloys Jestaedt, *Kataster der Stadt Fulda im 18. und 19. Jahrhundert,* Fuldaer Geschichtsverein publications 23–25.

154. *Wochenblatt der Provinz Fulda,* several years.

155. W. Lexis, *Die deutschen Universitäten,* Berlin, 1893 and 1904.

156. *Poggendorffs biographisch-literarisches Handwörterbuch,* vols. 3–6.

157. *Kurhessische Schulzeitung* (Kassel), several years.

158. Minutes, programs, and memorial publications of the Elector's *Gymnasium* in Fulda.

159. Announcements of courses and directories of the University of Marburg; address books 1868–1880ff.

160. Heinrich Hermelink and S. D. Kaehler, *Die Philipps-Universität Marburg 1527–1927,* Marburg, 1927.

161. Heinrich Doerbecker, *Marburg, seine Universität und deren Institute,* Marburg, 1927.

162. O. F. A. Schulze, *Zur Geschichte des Physikalischen Instituts der Universität Marburg,* in Ref. 160.

163. Franz Grundlach, *Catalogus Professorum Academiae Marburgensis,* Marburg, 1927.

164. Mannkopf, Address given on the occasion of the dedication of the new university building in Marburg, Marburg, 1879.

165. Paul Görlich, Fuldaer als Dozenten an der Marburger Universität, in Ref. 149 (1954).

166. L. Müller, *Marburger Studentenerinnerungen,* Marburg, 1906.

167. Georg Paul, *Marburger Studentenstreiche,* Berlin, 1887.

168. K. H. Müller, *Franz Melde,* Marburg, 1901.

169. Franz Melde, Selbstbiographie, in *Berichte des Freien Deutschen Hochstifts Frankfurt/Main,* Frankfurt, 1901.

170. P. Losch, Franz Emil Melde, *Bettelheims biographisches Jahrbuch* 6:338–339 (1904).

171. H. Kaemmerer, Ludwig Carius, in Ref. 143, 1875.

172. Almanacs, memorial publications, fraternity newspapers, annuals, and bluebooks of the fraternity Teutonia in Marburg.

173. Christian Buss, *Geschichte des Corps "Teutonia" at Marburg,* Marburg/Leipzig, 1907.

174. Philipp Braun, *Erinnerungen eines Marburger Teutonen.*

175. Directories of staff, students, and courses of lectures of the University of Berlin; address books 1869–1872ff.

176. Walter Bloch-Wunschmann, *Berlin.*

177. *Chronik der Technischen Hochschule Berlin.*

178. A. W. Dove, H. W. Dove, in Ref. 143 (1879).

179. A. W. Hofmann, Zur Erinnerung an Gustav Magnus, in Ref. 143 (1884).

180. Georg Cantor, Ernst Eduard Kummer, in Ref. 143 (1913).

181. Werner von Ziegfeld, Friedrich Harms, in Ref. 143 (1880).

182. Kurt Busse, Werner von Siemens, in Ref. 143 (1896).

183. Albert Ladenburg, Alphons Oppenheim, in Ref. 143 (1878).

184. W. Wedding, Adolf Slaby † [Obituary], *Elektrotech. Zeitschrift* 34:424, 429–430 (1913).

185. Walter König, Georg Hermann Quinkes Leben und Schaffen, *Naturwiss.* 12 (1924).

186. *Würzburger Universitätsalmanach* 1932–1933.

187. *Das neue Universitätsgebäude der Justus-Maximilians-Universität Würzburg,* memorial publication, Würzburg, 1897.

188. August Kundt, *Rede zur Feier des Stiftungstages der Universität Würzburg 1870*, Würzburg, 1870.

189. Programs of the Thomasschule in Leipzig, 1874–1877.

190. Franz Kemmerling, *Die Thomasschule zu Leipzig*, Leipzig, 1927.

200. Reinhard Hertz, *Album der Lehrer und Abiturienten der Thomasschule zu Leipzig 1832–1912*, Leipzig, 1912.

201. R. Knott, Wilhelm Hankel, in Ref. 143 (1900).

202. H. Reiger, Gustav Heinrich Wiedemann, in Ref. 143 (1898).

203. Reports of the *Rektor* and directories of staff, students, and courses of lectures of the University of Strasbourg; address books 1880–1882, 1895–1918ff.

204. A. E. Hoche, *Strassburg und seine Universität*, Berlin/Munich, 1939.

205. Otto Mayer, *Die Kaiser-Wilhelms-Universität Strassburg*, Berlin and Leipzig, 1922.

206. Gustav Andrich, *Die Kaiser-Wilhelms-Universität Strassburg in ihrer Bedeutung für die Wissenschaft*, Berlin and Leipzig, 1923.

207. Johannes Ficker, *Berichte über die Tätigkeit der Kriegsstelle der Universität Strassburg*, Strassbourg, 1914–1918.

208. Jean Pierhal, *Albert Schweitzer*, Munich, 1955.

209. Lists of courses and address books of the Technical University of Karlsruhe; city directories 1882–1885ff.

210. Edmund Sander, *Karlsruhe einst und jetzt*, Karlsruhe, 1911.

211. *Chroniken der Haupt- und Residenzstadt Karlsruhe*, 1882–1885.

212. *Festschriften der Technischen Hochschule Karlsruhe*, 1892, 1899, 1911, 1925, 1950.

213. Otto Lehmann, *Geschichte des Physikalischen Instituts der Technischen Hochschule Karlsruhe*, memorial publication, 1892.

214. *Karlsruher Zeitung*, 1883.

215. *Karlsruher Nachrichten*, 1884.

216. Annual reports and student newspapers of Polytechnischer Verein Karlsruhe, 1882–1885.

217. Der Altvater, home supplement of *Lahrer Zeitung*.

218. Lists of staff, students, and courses of the University of Tübingen; address books 1885–1895ff.

219. K. Fink, *Tübingen*, Tübingen, 1889.

220. T. Knapp and H. Kohler, *Die Universität Tübingen*, Düsseldorf, 1928.

221. *Festausgabe der Universität Tübingen*, 1893.

222. *Tübinger Chronik,* 1888, 1891, 1918; Festausgabe, 1927.

223. Paul von Grützner, *Die Dienstgesellschaft in Tübingen,* Tübingen, 1897.

224. *Berichte des Medizinisch-Naturwissenschaftlichen Vereins Tübingen,* 1885–1895.

E. HISTORY OF SCIENCE AND TECHNOLOGY

225. *Neue deutsche Biographie,* Berlin, 1955.

226. Karl Koppe, *Physik,* Essen, 1888.

227. Hermann von Helmholtz, *Vorträge und Reden,* Braunschweig, 1903.

228. A. Paalzow, Hermann von Helmholtz, in Ref. 143 (1894).

229. Helmut Unger, *Wilhelm Conrad Röntgen,* Hamburg, 1949.

230. Otto Glaser, *Wilhelm Conrad Röntgen und die Geschichte der Röntgenstrahlen,* Berlin, 1959.

231. Walther Gerlach, *Physik,* Frankfurt, 1960.

232. Henry Dufet, Sur la conductibilité électrique de la pyrite [On the electric conductivity of pyrite], *Comptes rendus* 81:628–631 (1875).

233. *Sitzungsberichte der Naturforschenden Gesellschaft zu Leipzig,* 1874–1877.

234. Werner Siemens, *Lebenserinnerungen,* Berlin, 1892, 1916; Munich, 1956; transl. W. C. Coupland, *Personal Recollections of Werner von Siemens,* London, 1893; 2nd ed., *Inventor and Entrepreneur: Recollections of Werner von Siemens,* London, 1966.

235. Georg Siemens, *Geschichte des Hauses Siemens,* Munich, 1947; transl. A. F. Rodger, *History of the House of Siemens,* Munich, 1957.

236. *50 Jahre Hartmann und Braun,* memorial publication, Frankfurt, 1932.

237. *Annalen der Physik und Chemie* (later *Annalen der Physik),* several years.

238. *Sitzungsberichte der Gesellschaft zur Beförderung der gesamten Naturwissenschaften in Marburg,* several years.

239. Armin Hermann, *Grosse Physiker,* Stuttgart, 1960.

240. Albert Gockel, *Über die Beziehungen der Peltierschen Wärme zum Nutzeffekt galvanischer Elemente* [On the relations of Peltier heat to the efficiency of galvanic elements], dissertation, University of Leipzig, 1885.

241. J. J. O'Neill, *Nikola Tesla,* New York, 1944.

242. *Bericht über die Verhandlungen der Sections-Sitzungen des Internationalen Elektrotechniker-Congresses zu Frankfurt am Main vom 7. bis 12. September 1891,* Frankfurt, 1892.

243. Otto Erhardt, *Über die Beziehungen der spezifischen Wärme und der Schmelzwärme bei hohen Temperaturen* [On the Relations of Specific Heat

and the Heat of Melting at High Temperatures], diss. University of Giessen, 1885.

244. August Schleiermacher, Über die Abhängigkeit der Wärmestrahlung von der Temperatur und das Stephan'sche Gesetz [On the dependence of the radiation of heat on temperature and Stephan's law], *Ann. Phys. Chem* (3)26:287–308 (1885).

245. *Sitzungsberichte der königlich preussischen Akademie der Wissenschaften in Berlin,* several years.

246. Heinrich Hertz, *Erinnerungen, Briefe, Tagebücher,* Leipzig, 1927; transl. (bilingual edition), *Memoirs, Letters, Diaries,* San Francisco, 1977.

247. Franz Wolf, Heinrich Hertz, in Ref. 143 (1897).

248. Heinrich Hertz, Über sehr schnelle electrische Schwingungen [On very rapid electric oscillation], *Ann. Phys. Chem* (3)31:421–448, 543–544 (1887).

249. Heinrich Hertz, Nachtrag [Addendum to Ref. 248], *ibid.*

250. *Electrotechnische Zeitschrift,* several years.

251. Ludwig Zehnder, Über Deformationsströme [On deformation currents] (correspondence on Ref. 35), *Ann. Phys. Chem.* (3)38:496 (1889).

252. Ludwig Zehnder, *Persönliche Erinnerungen an Röntgen und über die Entwicklung der Röntgenstrahlen,* Basel, 1938.

253. Mathias Cantor, *Über Capillaritätsconstanten* [On Capillary Constants], diss. University of Tübingen, 1892.

254. W. C. Röntgen, Vorläufige Mittheilung über eine neue Art von Strahlen [Preliminary communication on a new kind of rays], *Sitzungsber. physikal.-med. Ges. Würzburg* 132–141 (1895).

255. G. C. Schmidt, *Die Kathodenstrahlen,* Braunschweig, 1904.

256. Manfred von Ardenne, *Die Kathodenstrahlenröhre,* Berlin, 1933.

257. J. J. Thomson, Cathode rays, *Phil. Mag.* (5)44:293–316 (1897).

258. A. B. Macallum, Toronto meeting of the British Association, *Science* 5:251–252 (1897).

259. H. L. Callendar, Mathematics and physics at the British Association, *Science* 6:464–472 (1897).

260. Walther Gerlach, Fortschritte der Naturwissenschaften im 19. Jahrhundert, in *Propyläen-Weltgeschichte,* Berlin, 1960, vol. 8.

261. B. Walter, Über die Vorgänge im Induktionsapparat, *Ann. Phys. Chem.* 3(62):300–302 (1897).

262. W. H. Preece, Wireless telegraphy, *Electrician,* 1897; abstracted in *Elektrotech. Zeitschrift* 18:430–431 (1897).

263. K. Strecker, Über die Ausbreitung starker elektrischer Ströme in der Erdoberfläche [On the propagation of strong electric currents in the earth's surface], *ibid.* 17:106 (1896).

264. E. Rathenau, Telegraphie ohne metallische Leitung [Telegraphy without wire conductors], *ibid.* (1896).

265. J. N. Kane, *First Famous Facts,* New York, 1950.

266. Roger Louis, *Marconi—der Erfinder des Jahrhunderts,* Stuttgart, 1956.

267. F. Kiebitz, Nikola Tesla zum 75. Geburtstage [To Nikola Tesla on his 75th birthday], *Naturwiss.* 19:665–666 (1931).

268. Bruno Kuske, *100 Jahre Stollwerck-Geschichte* [One Hundred Years in the History of Stollwerck], Cologne, 1939.

269. Albert Hess, Reclamation [A claim], *Ann. Phys. Chem.* (3)64:623 (1898).

270. Albert Hess, Sur une application des rayons cathodiques à l'étude des champs magnétiques variables [On an application of cathode rays to the study of variable magnetic fields], *Comptes rendus* 119:57–58 (1894).

271. Public information office of Telefunken, several years.

272. *Die Naturwissenschaften,* several years.

273. Hans Gunther, *Pioniere der Radiotechnik,* Stuttgart, 1926.

274. *Jahrbuch der drahtlosen Telegraphie und Telephonie,* several years.

275. Nikola Tesla, *System of Transmission of Electrical Energy,* US Patent 645,576, 20 March 1900.

276. Wilheim Blohm, Die ersten Funkversuche in Cuxhaven, *Cuxhavener Zeitung,* 1937.

277. *Cuxhavener Tageblatt,* 1899–1901ff; esp. 27 September 1899.

278. City directories for Cuxhaven, 1899–1901.

279. Jonathan Zenneck, Aus der Kindheit der drahtlosen Telegraphie [On the early years of wireless telegraphy], *Telefunkenzeitung* (Berlin), 1922.

280. Arthur Wilke, Die Verbindung durch drahtlose Telegraphie zwischen Helgoland und Cuxhaven [Wireless-telegraphy connection between Helgoland and Cuxhaven], *Illustrirte Zeitung* (Berlin), 1901.

281. Hamburg *Courir,* 1900.

282. *Strassburger Post,* 1900.

283. Guglielmo Marconi, Nobel lecture, in *Les prix Nobel en 1909,* Stockholm: Norstedt & Soner, 1910; 24 pp.

284. *Physikal. Zeitschrift,* several years.

285. L. Mandelstam, *Bestimmung der Schwingungsdauer der oscillatorischen*

Condensatorladung [Determination of the Period of Oscillation of the Oscillatory Discharge], diss. University of Strasbourg, 1902.

286. Otto Jentsch, Der Konkurenzkampf auf dem Gebiete der Funkentelegraphie [The competitive struggle in the field of radiotelegraphy], Über Land und Meer supplement of *Deutsche illustrirte Zeitung* (Stuggart), 1902.

287. W. M. Varley, *Über den im Eisen durch schnell oszillierende Stromfelder induzierten Magnetismus* [On the Magnetism Induced in Iron by Rapidly Oscillating Current Fields], diss. University of Strasbourg, 1903.

288. Maximilian Rosen, Die Gesellschaft für drahtlose Telegraphie mbh, in *Die Führenden und ihr Werk*, Berlin, 1904.

289. N. Papalexi, *Ein Dynamometer für schnelle elektrische Schwingungen* [A Dynamometer for Rapid Electric Oscillations], diss. University of Leipzig, 1904.

290. Georg Rempp, *Die Dämpfung von Kondensatorkreisen mit Funkenstrecken* [Damping of Condensor Circuits Containing Spark Gaps], diss. University of Strasbourg, 1903.

291. Robert Feustel, *Über Kapillaritätsconstanten und ihre Bestimmung nach der Methode kleinster Blasen* [On Capillary Constants and Their Determination by the Method of Smallest Bubbles], diss. University of Strasbourg, 1903.

292. Gustav Aeckerlein, *Über die Zerstäubung galvanisch glühender Metalle* [On the Pulverization of Metals Brought to a Glow Galvanically], diss. University of Strasbourg, 1905.

293. Telefunken brochure, 1904.

294. *Allgemeine Zeitung* (Munich), 1905.

295. Max Dieckmann, *Über zeitliche Entladung von Schwingungen in Kondensatorkreisen* [On the Course of Oscillatory Discharges in Condensor Circuits], diss. University of Strasbourg, 1907.

296. Gustave Glage, *Experimentelle Untersuchungen im Resonanzinduktor* [Experimental Investigations of the Resonant Inductor], diss. University of Strasbourg, 1908.

297. *Les prix Nobel en 1909*, Stockholm, 1910.

298. Georg Graf Arco, Mögliches und Unmögliches in der drahtlosen Telegraphie [The possible and the impossible in wireless telegraphy], *Berliner Tageblatt*, 18 May 1904.

299. Hermann Rohmann, *Messung von Kapazitätsanderungen mit schnellen Schwingungen* [Measurement of Capacitive Changes with Rapid Oscillations], diss. University of Strasbourg, 1903.

300. E. Quäck, Neues über die Entwicklung der Grosstation Nauen [News on the development of the high-power transmitting station at Nauen], *Jahrbuch der drahtlosen Telegraphie and Telephonie* 13:333–342 (1918).

F. OTHER SOURCES

301. *Fliegende Blätter* (Munich), 1872–1878ff.

302. Papers of the Hesse State Archive, Marburg.

303. Documents and water colors in Ferdinand Braun's *Nachlass*, now in the Schloss Museum, Fulda.

304. Papers of the University of Würzburg.

305. Papers in the General Land Archives of Baden, Karlsruhe.

306. Papers of the University of Tübingen.

307. Documents in the company archives of Gebr. Stollwerck AG, Cologne.

308. Documents in the State Archive of Hamburg.

309. Emil von Borries, *Geschichte der Stadt Strassburg*, Strasbourg, 1909.

310. *Frankfurter Journal*, 9 December 1875.

311. *Intelligenzblatt der Stadt Frankfurt*, 7 December 1875.

312. *Staatszeitung und Herold* (New York), 23 March 1924.

313. *Physiker-Anekdoten*, Mosbach (Baden): Physik-Verlag, 1952.

314. *Fuldaer Tageblatt*, 6 June 1921.

315. *Der Spiegel* (Hamburg), No. 21, 1926.

316. *Hartmann und Braun 1899–1954*, unpublished manuscript.

317. *Deutsches Geschlechterbuch*, several volumes.

318. *Bolshaya Sovyetskaya Enciklopedia*.

319. *Daghens Nyheter* (Stockholm), 10 December 1909.

320. Communications from Konrad F. Braun, Kingston, RI.

321. Communications from Hildegard Stadler, Pforzheim.

322. Communications from Prof. Johannes Müller, Leipzig.

323. Communications from Paul Vitalli, Lahr.

324. Communications from Prof. Aloys Wuest, Freiburg.

325. Communications from Prof. Gustav Aeckerlein, Freiberg.

326. Communications from Hedwig Steiner, Münchberg.

327. Communications from Prof. Hermann Kehl, Munich.

328. Communications from Otto H. Stilling, Luxembourg.

329. Communications from Prof. Wenz, Fulda.

330. Communications from Walter Ungerer, Freiburg.

331. Communications from C. Morgenstern, Kronberg.

G. REFERENCES AND NOTES ADDED IN REVISED EDITION

332. B. F. Miessner, Detector for wireless apparatus, US Patent 1,104,065, 5 October 1910.

333. Arthur Schuster, On unilateral conductivity, *Phil. Mag.* (4)48:251–257 (1874).

334. W. G. Adams and R. E. Day, The action of light on selenium, *Proc. Roy. Soc.* 25:113–117 (1876).

335. E. W. von Siemens, Ueber den Einfluss der Beleuchtung auf die Leitungsfähigkeit des krystallinischen Selens, *Ann. Phys. Chem.* (2)156:334–335 (1875).

336. T. A. L. Du Moncel, Recherches sur la conductibilité électrique des corps médiocrement conducteurs et les phénomènes qui l'acompagnement, *Ann. Chim.* 10:194–271, 459–524 (1877).

337. Hermann von Helmholtz, Die Thermodynamik chemischer Vorgänge, *Sitzungsber. Akad. Wiss. Berlin* 22–39, 825–836 (1882); 647–665 (1883); see also vol. 124 of *Ostwalds Klassiker der exakten Wissenschaften.*

338. Franz Exner, Ueber galvanische Elemente, die nur aus Grundstoffen bestehen und über das Leitungsvermögen von Brom und Jod, *Ann. Phys. Chem.* (3)15:412–439 (1882).

339. Isaac Newton, *Principia mathematica,* London, 1687.

340. H. L. Le Châtelier, Sur un énoncé général des lois des équilibres chimiques, *Comptes rendus* 99:786–789 (1884); the principle is given in complete form in his Recherches expérimentales et théoriques sur les équilibres chimiques, *Ann. Mines et Carburants* (8)13:157–382 (1888).

341. H. R. Hertz, Ueber das Gleichgewicht schwimmender elastischer Platten, *Ann. Phys. Chem.* (3)22:449–445 (1884).

342. Henri Pellat, De la différence de potentiel entre électrodes et électrolytes, et de la polarisation, *Ann. Chim.* 19:556–574 (1890).

343. Theodor von Grotthus, Ueber die chemische Wirksamkeit des Lichts und der Electricität, und einen merkwürdigen neuen Gegensatz in der erstern, *Ann. Phys.* (1)61:50–64 (1819).

344. W. C. Röntgen, Ref. 234. The translation cited here is the version that appeared as "On a new form of radiation," in *Electrician* in London on 24 January 1896.

345. Julius Plücker, Ueber die Einwirkung des Magneten auf die electrische Entladung in verdünnten Gasen, *Ann. Phys. Chem.* (2)103:88–106, 151–157 (1858).

346. J. J. Fahie, *A History of Wireless Telegraphy*, Edinburgh, 1899.

347. Letter to Heinrich Huber, 3 December 1889; cited in Charles Süsskind, Hertz and the technological significance of electromagnetic waves, *Isis* 56:342–345 (1965).

348. Charles Süsskind, On the first use of the term 'radio,' *Proc. Inst. Radio Engrs* 50:326–327 (1962).

349. A. S. Popov, On the relation of metallic powders to electrical oscillations [in Russian], *Zh. Russ. Fiz.-khim. Obshchestva* (Physics, Part 1) 27:259–260 (1895).

350. A. S. Popov, Apparatus for the detection and recording of electrical oscillations [in Russian], *ibid.* 28:1–14 (1896). [Extracts appeared in a letter from Popov to *Electrician* 40:235 (1897).]

351. Charles Susskind, *Popov and the Beginnings of Radiotelegraphy*, San Francisco: San Francisco Press, 1962. [Reprinted from *Proc. Inst. Radio Engrs* 50:2036–2047; see also *Proc. Inst. El. and Electronics Engrs* 51:473–474, 959–960 (1963).]

352. Guglielmo Marconi, letter to *Electrical Rev.*, 14 February 1919, p. 179; cited in S. G. Sturmey, *The Economic Development of Radio*, London: Duckworth, 1958, pp. 19–20.

353. Adolf Slaby, *Die Funkentelegraphie*, Berlin, 1897.

354. B. W. Feddersen, Beiträge zur Kentniss des elektrischen Funkens, *Ann. Phys. Chem.* (2)103:69–88 (1858).

355. Nikola Tesla, Method of and apparatus for controlling mechanisms at a distance, US Patent 613,809, 1898.

356. The agreement was made in 1911; the case for Lodge and Muirhead was substantially strengthened by a privately printed pamphlet by S. P. Thompson, *Notes on Sir Oliver Lodge's Patent for Wireless Telegraphy*, London, 1911.

357. J. A. Fleming, *The Principles of Electric Wave Telegraphy*, London: Longmanns, 1906, p. 491.

358. S. G. Brown, Directional wireless, British Patent 14,449, 1899.

359. Charles Süsskind, Guglielmo Marconi (1874–1937), *Endeavour* 33:67–72 (1974).

360. Emil Rathenau (1838–1915), founder of AEG's predecessor (the German Edison Co.), was the father of Walther Rathenau (1867–1922), who succeeded him as AEG's chief executive in 1915 and later, until he was assassinated by reactionaries, served as Germany's foreign minister.

361. Nikola Tesla, The problem of increasing human energy, *Century Mag.* 60:175–211 (1900).

362. S. G. Brown's 1899 patent (Ref. 358) is the earliest publication in radio detection finding; the first practical apparatus ('radiogoniometer') is generally ascribed to Ettore Bellini and Alessandro Tossi, Wireless telegraphy, British Patent 21,299, 25 September 1907, also described in their paper, A directive system of wireless telegraphy, *Phil. Mag.* (6)16:638–657 (1908).

363. The "singing arc" method had been proposed by William Duddel in British Patent 21,629, 29 November 1900, and in an article in *Electrician* 46:269, 310 (1900).

364. The suggestion that the award was made jointly "to the surprise of the whole world" and that Braun "charmingly apologized for being there at all" are made in the biography by Degna Marconi, *My Father, Marconi,* New York: McGraw–Hill, 1962, pp. 182–190. This is a view that stems from the great world renown of Marconi, who was doubtless better known that Braun to the general public, though not necessarily to the selection committee of Swedish physicists. Quite apart from his many scientific contributions, Braun was mentioned in every textbook on wireless telegraphy written by his contemporaries; for instance, those of Slaby (Ref. 353), Fleming (Ref. 357), and of course his own monograph (Ref. 68). The pioneering American radio scientist G. W. Pierce commended Braun five times in his text *Principles of Wireless Telegraphy* (New York: McGraw–Hill, 1910): for making use of inductive coupling to transmitter antennas (pp. 101–103), for suggesting the artificial ground (p. 121), for the usefulness of his cathode-ray oscilloscope (pp. 151 and 181–182), and for originating use of the parabolic and the phased wire arrays as directional antennas (pp. 296–297).

365. In an historical account written at the Bell Laboratories soon after the transistor was discovered, coauthored by one of the discoverers (W. H. Brattain, who shared the 1956 Nobel Prize in physics with William Shockley and John Bardeen), Braun is given full credit for the discovery that the resistance of certain materials depends on the magnitude and sign of the applied voltage (the rectifier effect): G. L. Pearson and W. H. Brattain, History of semiconductor research, *Proc. Inst. Radio Engrs.* 43:1794–1806 (1955).

366. Charles Süsskind, The early history of electronics, *Inst. El. and Electronics Engrs. Spectrum* 5(8):90–98, (12):57–60 (1968); 6(4):69–74, (8): 66–70 (1969); 7(4):78–83 (see esp. p. 83), (9):76–79 (1970).

367. See the *New York Times* and other 1915 newspapers; also, L. S. Howeth, *History of Communications-Electronics in the United States Navy,* Washington: Government Printing Office, 1963; Charles Süsskind, Ferdinand Braun: Forgotten Forefather, in L. and C. Marton, Eds., *Advances in Electronics and Electron Physics* 50:241–260, New York: Academic Press (1980).

NAME INDEX

NAME INDEX

Thiele, Johannes, 175, 176, 216
Thompson, S. P., 94
Thomson, Sir Joseph John, 180
Thomson, Sir William, 40, 60, 94
Tietz, Martin, 148
Töpler, August, 59
Tyndall, John, 15

Ungerer, Arnold, 193, 194, 196
Ungerer, Walter, 193–194

Varley, William Mansergh, 152
Voller, August, 98, 140

Waitz, Karl, 73, 74
Waldemar, Prince, 171
Walter, Berhard, 98
Warburg, Emil, 35
Weber, Wilhelm, 35, 101
Wedel, Count von, 171, 187
Wehnelt, Arthur Rudolf Berthold, 99, 188
Wehner, Josef, 119, 140
Wellstein, Josef, 82, 86
Wiechert, Emil, 100
Wiedemann, Gustav, x, 15, 34, 39, 44–45, 48, 97
Wiegand, Karl H. von, 209
Wien, Max, 144, 149, 177, 231, 238
Wilhelm II, 70, 106, 148–149, 153, 155, 163–165
Winkelmann, Professor, 79, 158
Wrede, J., 109
Wuest, Alois, 174
Wüllner, Adolf, 15

Zeemann, Pieter, 180
Zehnder, Ludwig, 71, 72
Zenneck, Jonathan, 19, 66, 73–74, 82–83, 85, 91, 92, 112, 130–131, 133, 134, 135, 136, 140, 144, 145, 156, 157, 163, 168–169, 173, 193, 196, 202, 204–205, 207, 208, 210, 212, 215, 220, 222, 223
Zeppelin, Ferdinand von, 173, 175, 196, 197
Zimmer, Dr., 140
Zobel, Albert, 101, 108, 110, 117, 119, 121, 123, 125, 126, 137, 140

SUBJECT INDEX

Capillarity, 24, 43, 158
Cathode-ray oscilloscope, 89–93,
  95–100, 145–146, 152, 161, 222
Central Bureau for Electricity,
  Karlsruhe, 59
Central Office for Meteorology
  and Hydrography, Grand
  Duchy of Baden, 58–59
Central Radio Laboratory, Lenin-
  grad, 197
*Centralzeitung für Optik und
  Mechanik*, 52
Chemisches Central-Blatt, 7–8
Cinematograph, 86, 109
"Circuit to Strengthen Electrical
  Waves, A" (Braun transmitter
  patent), 133
Contemporary Culture, 211
"Coupled Condensor Circuits
  with Very Short Spark Gaps"
  (Riegger), 194
Crimped-wire electrode, 29, 30
"Criterion for Determining the
  Cohesion of a Surface Layer
  and a Note on the Vapor Pres-
  sure of Such Layers, A," 112
Crystal detector, 131–132, 163,
  196
Crystallography, Braun textbook
  on, 6, 46
"Current Waveforms," 91
Cuxhaven, radiotelegraphic exper-
  iments at, 123–124, 125, 126,
  127, 128–131, 134–137, 196,
  208, 232

Daghens Nyheter, 184–185
"Damping of Condensers in
  Spark-Gap Circuits" (Rempp),
  157
"Damping and Energy Efficiency
  of Several Transmission Con-
  figurations in Radiotelegraphy"
  (Brandes), 174
De Forest Wireless Telegraph
  Co., 188
"Deformation currents," 70–72,
  74
"Determination of the Period of
  Oscillation of an Oscillatory

Condenser Discharge" (Man-
  delstam), 145
Deutsche Staatszeitung, 175, 209,
  220
Deutschland, 137
"Directed Wireless Telegraphy,"
  170
Dissertation, Braun, 17–20, 35,
  56, 158
"Do Cathode Rays Exhibit Uni-
  polar Rotation?," 112
"Dynamometer for Rapid Electric
  Oscillations, A" (Papalexi), 157

Earth's interior, measuring tem-
  perature of, 74
Eiffel Tower, antenna experi-
  ments, 191–192, 193, 195
Elbe II, 133, 137
Electrical engineering, 49–51
"Electrical Oscillations and Wire-
  less Telegraphy" (Braun Nobel
  lecture), 184, 228–249
Electrician, The, 170, 195
Electrolytic conduction, 24
Electromagnetic Field, The (Cohn),
  82
*Electromagnetic Oscillations and
  Wireless Telegraphy* (Zenneck),
  145
Electron microscope, 161
Electrostenolysis, 76
Electrotechnische Zeitschrift, 78, 99,
  106, 142, 148, 191
Elektrichestvo, 86
Elektrizitäts-AG, 134
Elka Park, NY, 209, 212–213,
  215
"Experimental Investigations of
  Braun's Dispersion Grating"
  (A. Ungerer), 194
"Experimental Investigations of
  the Resonant Inductor" (Glage),
  178
"Experiments to Demonstrate
  Electrical Surface Conductivity
  in a Preferred Direction," 87
"Experiments on Departures from
  Ohm's Law in Metallic Con-
  ductors," 33

"On the Physical Interpretation of Thermoelectricity," 80
"On the Processes in an Induction Machine" (Walter), 97
"On the Pulverization of Dynamically Heated Metals" (Äckerein), 162
"On Rational Transmitter Circuits in Wireless Telegraphy," 142
"On the Relations of Peltier Heat and the Efficiency of Galvanic Elements" (Gockel), 51
"On the Relations of Specific Heat and the Heat of Melting at High Temperatures" (Ehrhardt), 51
On the Sensations of Tone (Helmholtz), 17
"On Some Alternative Connections for Wireless Telegraphy," 143
"On Spherical Functions," 40
"On the Substitution of Closed Current Paths for Open Paths in Wireless Telegraphy," 193
"On the Thermoelectricity of Molten Metals," 54
On the Transmission of Electrical Energy, Particularly Alternating Current, 77
"On the Velocity of Propagation of Wave Motions and Single Pulses in Membranes Bounded by Water on One or Both Sides, with Special Consideration of Conditions of the Ear" (Wuest), 174
Optics, 158–159

Palatia, 168
Patria, 135
Phased antennas, 167–168
"Philosopher Guardians, The," 23
Photovoltaic effect, 68
"Physics for Women," 213–215
Physikalische Zeitschrift, 142, 156
Polytechnic Union, Karlsruhe, 56

Pravda, 146
"Pupin coils," 220

Radiation experiments, 88–89
Radioactivity, 166
Radio science, 143–144
Radiotelegraphy, 105–108, 141–147, 157, 158
Atlantic bridged, 147
aviation, 196–197
Cuxhaven, experiments, 123–124, 125, 126, 127, 128–131, 134–137, 196, 208, 232
Friedrichshaven, experiments, 196–197
Germany, 108–122, 131–138
Great Britain, 195–197
United States, 189–190
World War I, 200
Reclamation (Hess), 98
Rectifier effect, 28–29, 44, 45, 131
Rock salt, electric properties of, 68–69
Royal Bavarian Academy of Sciences, Munich, 63
Royal Prussian Academy of Sciences, 56
Royal Swedish Academy of Sciences, 179, 180–181, 186

Sayville, Long Island, 190, 202, 203, 204, 206, 208, 215–216
Schuckert & Co. (Elektrizitäts-AG), 47, 134–135, 137, 138, 139
Science, 94
Secret societies, 8–9
Siemens Co., 138, 139–140, 143, 189
merger with Telebraun, 138–140
Siemens & Halske, 50, 109, 139, 140, 152, 153
Silvana, 130, 133, 134, 195
Skin effect, ac and, 108, 109
Society for the Advancement of Natural Sciences, Marburg, 39

Wireless Telegraph and Signal
    Co., 106, 109, 127
"Wireless telephony," 211
Woche, Die, 162
World War I, 195, 198–218
Würzburg Physics Institute, 25–26
Würzburg, University of, 21–26

X ray, 84–88

Zeitschrift für physikalische Chemie,
    63